T0297925

# LECTURES ON
# ARAKELOV GEOMETRY

*Already published*

# LECTURES ON
# ARAKELOV GEOMETRY

C. SOULÉ

*CNRS,*
*Institut des Hautes Études Scientifiques*

written with

D. ABRAMOVICH, J.-F. BURNOL & J. KRAMER

Published by the Press Syndicate of the University of Cambridge
The Pitt Building, Trumpington Street, Cambridge CB2 1RP
40 West 20th Street, New York, NY 10011-4211, USA
10 Stamford Road, Melbourne, Victoria 3166, Australia

© Cambridge University Press 1992

First published 1992

First paperback 1994

*Library of Congress cataloguing in publication data available*

*A catalogue record for this book is available from the British Library*


ISBN 0 521 41669 8   hardback

ISBN 0 521 47709 3   paperback


Transferred to digital printing 2003

# Contents

vi

# Foreword

In the first semester of 1989, I was invited to give a graduate course in the Mathematics Department of Harvard University. Notes were written by D. Abramovich, J.-F. Burnol and J. Kramer, and typed in TeX by L. Schlesinger in that department. They were revised and expanded during the next two years. My coauthors corrected and completed the contents of the book, especially by providing some of the proofs in Chapter VI. I also benefited from the advice of E. Getzler, who let me use a preliminary draft of his book with N. Berline and M. Vergne [BGV]. M. Hindry and A. J. Scholl read the manuscript carefully, and offered useful comments.

Without the efficient and generous help of all these people, these notes would not exist. I thank them very warmly, as well as the Harvard Mathematics Department for its hospitality.

C. Soulé.

The material presented in this book is joint work with H. GILLET. Chapters VI and VII describe joint work with H. GILLET and J.-M. BISMUT.

# 0

---

# Introduction

## 1. The method of infinite descent

**1.1**  One way of proving the irrationality of the square root of 2 is the following one, known as the *infinite descent* method [I]. Assume that $x$ and $y$ are non-zero positive integers such that

$$(1) \qquad\qquad x^2 = 2y^2.$$

Then $x^2$ is even, hence $x$ is even,

$$(2) \qquad\qquad x = 2z,$$

say. Substituting in (1) and dividing by two we get

$$(3) \qquad\qquad y^2 = 2z^2.$$

In other words, we have a new solution of (1). But this solution $(y, z)$ is *smaller* than the previous one. For instance, the squared norm $y^2 + z^2$ of the vector $(y, z)$ is smaller than $x^2 + y^2$.

We can repeat the argument starting from the solution $(y, z)$ to get a new solution $(z, t)$, and so on. Since a bounded set of integers is finite, this "descent" must end. We get a solution of (1) with either $x$ or $y$ equal to zero. But then, by (2), the original solution must also be the trivial one; a contradiction.

**1.2**  One may wonder whether this method of infinite descent, due to Fermat, can solve all diophantine equations, i.e., systems of polynomial equations with integral coefficients and integral unknowns. It is very appropriate for showing that a diophantine equation has no nontrivial solution; for instance, $x^4 + y^4 = z^2$ or $x^3 + y^3 = z^3$ [W1].

It is also a good method for obtaining infinitely many solutions. One starts with a few small solutions; then, by running the descent argument backwards, these generate infinitely many others. This applies to Pell's equation and to the proof of the Mordell–Weil theorem.

But the argument we just made suggests that, when descent applies, there are either zero or infinitely many nontrivial solutions. Indeed, if there are finitely many solutions, when trying to show there are no more, we would have to be sure that the descent process avoids the nontrivial ones!

**1.3**    To overcome this difficulty, we describe the method of infinite descent in other terms. Forgetting the (fortuitous?) fact that equation (3) repeats equation (1), we may say that *infinite descent is a combination of congruence and height (= size) arguments*. In that sense, one may hope for a *geometry* which, instead of using these two kinds of arguments one after the other, would involve them simultaneously. Such a static version of infinite descent might prove finiteness theorems.

**1.4**    The Grothendieck theory of *schemes* gives an adequate geometric generalization of the congruence arguments in diophantine equations. Indeed, a scheme $X$ over $\mathbf{Z}$ is viewed as a family of varieties over Spec $\mathbf{Z}$, the fiber at a prime $p$ being the reduction of $X$ modulo $p$.

Given such a variety $X$, if we want to control the *height* of its points we have to consider the complex variety $X(\mathbf{C})$ (we assume it is smooth) from the point of view of *hermitian complex geometry*. This means that we endow holomorphic vector bundles on $X(\mathbf{C})$ with smooth hermitian metrics.

*Arakelov geometry* [A1][A2] is a combination of schemes and hermitian complex geometry. Its main achievement today is the proof of the Mordell conjecture [F1][V2][Bo]: a smooth projective curve of genus greater than one has only finitely many rational points.

**1.5**    As we said earlier, Arakelov geometry is a *static* generalization of infinite descent. For instance, when doing intersection theory on $X$ (see Chapter I below) one is not allowed to move the cycles; no analog of Chow's Moving Lemma is known over $\mathbf{Z}$. A more dynamic approach would be an *adelic* variant of Arakelov geometry. The main object of study in this theory would be a smooth variety $V$ over $\mathbb{Q}$, and vector bundles on $V$ equipped with metrics at archimedean places, and $p$-adic analogs of these at finite places. Such an *adelic geometry* is still to be built.

## 2. The analogy between function fields and number fields

**2.1**  Several authors, including A. Weil [W2], have emphasized the analogy between a number field, i.e., a finite extension of $\mathbb{Q}$, and the field $\mathbb{C}(S)$ of meromorphic functions on a smooth complete curve $S$.

For instance, for any function $f \in \mathbb{C}(S)$, $f \neq 0$, and any point $x \in S$, denote by $v_x(f) \in \mathbb{Z}$ the valuation of $f$ at $x$, i.e., the order of vanishing of $f$ at $x$ or minus the order of the pole of $f$ at $x$. From the Cauchy *residue formula* we get

$$(4) \qquad \sum_{x \in S} v_x(f) = \sum_{x \in S} \operatorname{Res}_x \left( \frac{df}{f} \right) = 0,$$

where $\operatorname{Res}_x$ denotes the residue at $x$ of differential forms.

When $f \in \mathbb{Q}^*$ is a rational number we have the *product formula*

$$(5) \qquad |f| = \prod_p p^{v_p(f)},$$

where $p$ runs over all integral primes and $v_p(f) \in \mathbb{Z}$ is the $p$-adic valuation of $f$. If we define

$$(6) \qquad v_\infty(f) = -\log|f| \in \mathbb{R},$$

we may rewrite (5) as

$$(7) \qquad \sum_p v_p(f) \log(p) + v_\infty(f) = 0,$$

an analog for $\mathbb{Q}$ of equation (4) for $\mathbb{C}(S)$.

From this example we see that, in this analogy, the complete curve $S$ is analogous to the affine scheme $\operatorname{Spec}\mathbb{Z}$ to which is added a point at infinity (at this point the archimedean norm is used instead of discrete valuations). This fits with the view expressed above that algebraic geometry has to be completed by hermitian complex geometry.

**2.2**  In general, let $X$ be an *arithmetic variety*. By this we mean a regular scheme, projective and flat over $\mathbb{Z}$.

In other words, we consider a system of polynomial equations

$$(8) \qquad f_1(x_0, \cdots, x_N) = f_2(x_0, \cdots, x_N) = \cdots = f_k(x_0, \cdots, x_N) = 0,$$

where $f_1, \cdots, f_k \in \mathbb{Z}[X_0, \cdots, X_N]$ are homogeneous polynomials with integral coefficients. These define the projective scheme $X = \mathrm{Proj}(S)$, where $S$ is the quotient of $\mathbb{Z}[X_0, \cdots, X_N]$ by the ideal generated by $f_1, \cdots, f_k$. The points of $X$ are those homogeneous prime ideals $\mathcal{P}$ in $S$ which do not contain the augmentation ideal ([H], II.2). The map $f : X \to \mathrm{Spec}\,\mathbb{Z}$ maps $\mathcal{P}$ to $\mathcal{P} \cap \mathbb{Z}$. The fiber of $f$ over a prime integer (special fiber) is the variety $f^{-1}(p\mathbb{Z}) = X/p = \mathrm{Proj}(S/pS)$ over the field with $p$ elements. The generic fiber is $f^{-1}((0)) = X_{\mathbb{Q}} = \mathrm{Proj}(S \otimes_{\mathbb{Z}} \mathbb{Q})$. We assume that $X$ is regular and that $f$ is flat, i.e. $S$ is torsion free. It follows that $X/p$ is smooth, except for finitely many values of $p$, like $q$ in Figure 1, where it may not even be reduced.

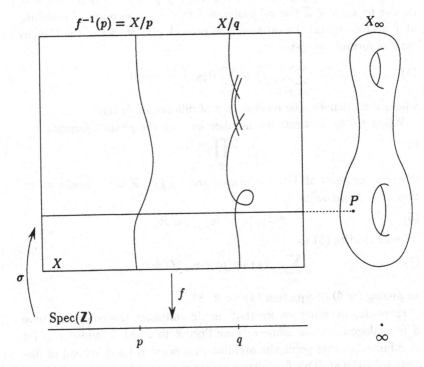

Figure 1: An arithmetic variety

In the same way that we completed $\mathrm{Spec}\,\mathbb{Z}$ by adding a point $\infty$ to it, we "complete" the family $X$ of varieties over $\mathrm{Spec}\,\mathbb{Z}$ by adding to it the complex variety $X_{\infty} = X(\mathbb{C})$, i.e. the set of complex solutions of (8), viewed as the fiber at infinity. We think of the whole family as analogous to a complete smooth complex manifold $Y$ fibered over a smooth complete curve $S$ via a flat proper map $f : Y \to S$, and we visualize the situation as in Figure 1. If the fibers of $f$ have dimension one, $X$ has Krull dimension two (we call it an *arithmetic surface*) and $X_\infty$ is a complex curve (a *Riemann surface*). Notice that an integral solution of (8) is a (rational) point $P$ in $X(\mathbb{Z}) = X(\mathbb{Q}) \subset X(\mathbb{C})$, i.e. a section $\sigma$ of $f$.

The need to control the size of these points $P$ leads us to take as our main object of study an algebraic vector bundle $E$ on $X$, endowed with a smooth hermitian metric $h$ on the corresponding holomorphic vector bundle $E_\infty$ on $X_\infty$ (we further assume that $h$ is invariant under the complex conjugation $F_\infty$ on $X_\infty$). The pair $\overline{E} = (E, h)$ will be called an *hermitian vector bundle* on $X$.

## 3. The contents of this book

**3.1**   Let $X$ be an arithmetic variety and $\overline{E}$ an hermitian vector bundle on $X$. We shall attach to $\overline{E}$ *characteristic classes* with values in *arithmetic Chow groups*.

More specifically, an *arithmetic cycle* is a pair $(Z, g)$ consisting of an algebraic cycle on $X$, i.e. a finite sum $\sum_\alpha n_\alpha Z_\alpha$, $n_\alpha \in \mathbb{Z}$, where $Z_\alpha$ is a closed irreducible subscheme of $X$, of fixed codimension $p$, say, and a *Green current* $g$ for $Z$. By this we mean that $g$ is a real current on $X_\infty$ which satisfies $F_\infty^*(g) = (-1)^{p-1} g$ and

$$(9) \qquad\qquad dd^c g + \delta_Z = \omega,$$

where $\omega$ is (the current attached to) a smooth form on $X_\infty$, and $\delta_Z$ is the current given by integration on $Z_\infty$:

$$(10) \qquad\qquad \delta_Z(\eta) = \sum_\alpha n_\alpha \int_{Z_\alpha(\mathbb{C})} \eta,$$

for any smooth form $\eta$ of appropriate degree.

The arithmetic Chow group $\widehat{CH}^p(X)$ is the abelian group of arithmetic cycles, modulo the subgroup generated by pairs $(0, \partial u + \overline{\partial} v)$ and $(\operatorname{div} f, -\log|f|^2)$, where $u$ and $v$ are arbitrary currents of the appropriate degree and $\operatorname{div} f$ is the divisor of a non-zero rational function $f$ on some irreducible closed subscheme of codimension $p - 1$ in $X$.

**3.2**   In Chapter III we study the groups $\widehat{CH}^p(X)$, showing that they have functoriality properties and a graded product structure, at least after tensoring them by $\mathbb{Q}$. To prove these facts is rather difficult, for two reasons.

First, the intersection theory on a general regular scheme such as $X$ cannot be defined in the usual way, since no Moving Lemma is available. We remedy this in Chapter I by using algebraic $K$-theory and Adams operations as in [GS1]. In particular, we give a proof of the vanishing of Serre's intersection multiplicities of two modules on a regular local ring, when the sum of the codimensions of their supports exceeds the dimension of the ring.

A second difficulty is that, given two arithmetic cycles $(Z, g)$ and $(Z', g')$, we need a Green current for their intersection. The formula

$$g'' = \omega g' + g\delta_{Z'},$$

where $\omega$ is defined as in (9), is formally satisfactory, but involves a product of currents $g\delta_{Z'}$. To make sense of it in general we need to show that we can take for $g$ a smooth form on $X_\infty - Z_\infty$, of *logarithmic type* along $Z_\infty$. This is done in Chapter II.

**3.3**    After having set up arithmetic intersection theory, we define in Chapter IV characteristic classes for hermitian vector bundles $\overline{E}$ on $X$. For instance, we get a Chern character class

$$(11) \qquad \widehat{\mathrm{ch}}(\overline{E}) \in \bigoplus_{p \geq 0} \widehat{CH}^p(X) \otimes_{\mathbf{Z}} \mathbf{Q}.$$

This class satisfies the usual axiomatic properties of a Chern character. But it does depend on the choice of a metric on $E$. Furthermore it is not additive for arbitrary exact sequences; it is additive, however, on orthogonal direct sums. Its failure to be additive on exact sequences is given by a *secondary characteristic class* first introduced by Bott and Chern [BC]. Similar results hold for the Chern classes $\widehat{c}_n(\overline{E})$ and the Todd class $\widehat{Td}(\overline{E})$.

**3.4**    Our next construction is some *direct image* map for hermitian vector bundles. Let $f : X \to Y$ be a proper flat map between arithmetic varieties, smooth on the generic fiber $X_{\mathbf{Q}}$. According to [KM], there is a canonical line bundle $\lambda(E)$ on $Y$ whose fiber at every point $y \in Y$ is the determinant of the cohomology of $X_y = f^{-1}(y)$ with coefficients in $E$:

$$(12) \qquad \lambda(E)_y = \bigotimes_{q \geq 0} \Lambda^{\max}(H^q(X_y, E))^{(-1)^q},$$

where $\Lambda^{\max}$ denotes the maximal exterior power, and $L^{-1}$ the dual of a line bundle $L$.

To get a metric on $\lambda(E)$ let us fix a Kähler metric on $X_\infty$, hence on each fiber $X_y$, $y \in Y_\infty$. According to Quillen [Q2] we may then a smooth metric $h_Q$ on $\lambda(E)_\infty$ by multiplying its $L^2$-metric (given by integration along the fibers) by the square of the Ray–Singer *analytic torsion* [RS]:

(13) $$h_Q = h_{L^2} \cdot \exp(T(E)),$$

with

(14) $$T(E) = \sum_{q \geq 0} (-1)^{q+1} q\, \zeta_q'(0).$$

Here $\zeta_q'(0)$ is the derivative at the origin of the zeta function $\zeta_q(s)$, $s \in \mathbb{C}$, of the Laplace operator $\Delta^q = \overline{\partial}\overline{\partial}^* + \overline{\partial}^*\overline{\partial}$ acting upon forms of type $(0, q)$ on $X_y$, $y \in Y_\infty$, with coefficients in $E_\infty$.

In Chapters V and VI we study $\zeta_q(s)$ and the Quillen metric. Following [BGS1] we show that $h_Q$ is smooth and we compute in Chapter VII the curvature on $Y_\infty$ of the hermitian line bundle $\lambda(E)_Q = (\lambda(E), h_Q)$. It is given by a Riemann–Roch–Grothendieck formula at the level of forms.

**3.5** When combining the above results with the Riemann–Roch–Grothendieck theorem for algebraic Chow groups, we get in Chapter VIII a Riemann–Roch–Grothendieck theorem for arithmetic Chow groups. Given a proper map $f : X \to Y$ between arithmetic varieties, smooth on $X_\mathbb{Q}$, and an hermitian vector bundle $\overline{E}$ on $X$, this theorem states that

(15) $$\delta(E) = \widehat{c}_1(\lambda(E)_Q) - f_*(\widehat{ch}(\overline{E})\widehat{Td}(f))^{(1)}$$

depends only on the class of $E$ in the Grothendieck group $K_0(X_\mathbb{Q})$ of the generic fiber of $X$; here $\alpha^{(1)}$ is the degree one component of $\alpha \in \bigoplus_{p \geq 0} \widehat{CH}^p(X) \otimes_\mathbb{Z} \mathbb{Q}$.

**3.6** An application of this is the following existence theorem of small sections for powers of ample line bundles [GS4]. Let $\overline{L}$ be an hermitian line bundle on some arithmetic variety $X$, of relative dimension $d$ over $\operatorname{Spec} \mathbb{Z}$. Assume that $L$ is ample, the metric on $L_\infty$ is positive, and the arithmetic self-intersection $\overline{L}^{d+1} \in \mathbb{R}$ of $\overline{L}$ is positive. Let $\overline{E}$ be any holomorphic vector bundle on $X$, and $r$ the rank of $E$. Call $h^0(X, \overline{E} \otimes \overline{L}^n) \in \mathbb{R}$ the logarithm of the number of sections $s \in H^0(X, E \otimes L^n)$ of $E \otimes L^n$ such that

(16) $$\|s(x)\| \leq 1 \quad \text{for every } x \in X_\infty;$$

compare with (6).

Then, as $n$ goes to infinity, we have

(17) $$h^0(X, \overline{E} \otimes \overline{L}^n) \geq \frac{r}{(d+1)!} \overline{L}^{d+1} n^{d+1} + O(n^d \log n).$$

The proof of this result combines the arithmetic (relative) Riemann–Roch–Grothendieck theorem for the map $X \to \operatorname{Spec} \mathbb{Z}$ with the Minkowski theorem for the lattice $H^0(X, E \otimes L^n)$, endowed with the appropriate metrics. This fits well with the view of Weil [W2] that Minkowski's theorem is an arithmetic analog of the (absolute) Riemann–Roch theorem for complex curves; see also [GS6]. Notice that the base point over which such a curve is defined has no obvious arithmetic counterpart!

The proof of (17) was used by Vojta in one of the steps of his proof of Mordell's conjecture. We refer the reader to his paper [V2], and to Faltings' paper [F3] for the use of arithmetic intersection theory in the study of rational points on abelian varieties.

**3.7**  Finally, a word of warning. Several proofs in this book are only sketched. Furthermore, we shall quote without proofs results from algebraic $K$-theory ([Q1], [S1], [GS1]) and the family index theorem ([B1], [BV], [BGV]). Generally speaking, we assume more knowledge of algebra, especially in Chapter I, than of differential geometry, but it might help to consult other books for the basic material, for instance [BGV] when reading Chapters V, VI and VII.

The book contains several remarks and open problems. These range from precise assertions and references to vague conjectures. We have not censored them too much, rather hoping that they will stimulate the reader's own research.

# I

---

# Intersection Theory on Regular Schemes

In this chapter, we shall follow [GS1] and define an intersection theory on an arbitrary regular noetherian finite-dimensional scheme $X$, i.e. a graded pairing between the Chow groups $CH^p(X)$ of cycles of codimension $p$ on $X$, modulo linear equivalence. However, in general, this pairing is defined only up to torsion.

When $X$ is of finite type over a field, the usual method to get an intersection theory for cycles on $X$ is to use the Moving Lemma [RJ], which asserts that, given two cycles, one can change one of them by linear equivalence and make their intersection proper. Unfortunately this Lemma is not known on a general base. When $X$ is smooth over a Dedekind ring, Fulton's method of the normal cone can be applied instead of the Moving Lemma [Fu]. But in general no geometric method is available (see however [RP] and [KT] for an extension of Fulton's method, up to torsion).

The tool we shall be using here is an isomorphism between $CH^p(X)_{\mathbb{Q}}$ and $K_0(X)^{(p)}$, the weight $p$-part, for the Adams operations, of the Grothendieck $K$-group of locally free coherent $\mathcal{O}_X$-modules. The pairing between $K_0(X)^{(p)}$ and $K_0(X)^{(q)}$ is then just given by the tensor product of $\mathcal{O}_X$-modules.

The plan of this chapter is as follows. In §1 and §2 we state the main results (see Theorem 2). In §3 we introduce Grothendieck groups. In Theorem 3(i) we state that their filtration by codimension is multiplicative up to torsion, and in Theorem 3(ii) we compare it with the Chow groups. In Corollary 1 we deduce from Theorem 3 a conjecture of Serre on the vanishing of some intersection multiplicities; another proof was given by Roberts [RP], using Fulton's theory. In §4 we define $\lambda$-rings and we state the existence of such a structure on the Grothendieck groups (with supports). This result is proved in §5. In §6 we prove Theorem 3(i), thus completing the proof of Serre's conjecture. Finally, we describe how the $\lambda$-ring structure on Quillen higher $K$-theory [Kr] [S1] leads to the comparison of Chow groups and Grothendieck groups stated in Theorem 3(ii). However, at this point, we do not give any details.

We shall use the following convention (valid for the whole book): given an abelian group $A$, we denote by $A_{\mathbb{Q}}$ the vector space $A \otimes_{\mathbf{Z}} \mathbb{Q}$.

## 1. Length and order

**1.1**     Let $R$ be a noetherian ring and $M$ a finitely generated $R$-module. There exists a chain of submodules

$$(1) \qquad\qquad M = M_0 \supset M_1 \supset \cdots \supset M_\ell = 0$$

with $M_{i-1}/M_i \cong R/\wp_i$, where $\wp_i$ is a prime ideal of $R$ [Se2].

**Definition 1**  *M is said to have finite length, if all $\wp_i$ occurring in (1) are maximal ideals.*

A module $M$ has finite length if and only if its support Supp $M = \{\wp \in \mathrm{Spec} R : M_\wp = M \otimes_R R_\wp \neq 0\}$ consists of maximal ideals. If $M$ has finite length, it can be shown that any two chains (1) have the same length; we denote it by $\ell_R(M)$ and call it the *length* of $M$. The function $\ell_R(\cdot)$ is additive on exact sequences.

**1.2**     Let now $R$ be a one-dimensional integral domain and $K$ its fraction field.

**Definition 2**  *For $f \in K^*, f = a \cdot b^{-1}$ with $a, b \in R$ we put*

$$\mathrm{ord}_R(f) := \ell_R(R/aR) - \ell_R(R/bR)$$

*and call it the order of $f$.*

Then $\mathrm{ord}_R : K^* \to \mathbf{Z}$ is a homomorphism from the multiplicative group $K^*$ to the additive group $\mathbf{Z}$. If $R$ is a one-dimensional regular local ring, then the order of any $f \in R$ coincides with the valuation of $f$; note that $R$ is then a discrete valuation ring.

## 2. Chow Groups

Let $X$ be a noetherian and separated scheme of dimension $d$, an assumption which will be made during the whole of Chapter I. For any integer $p \geq 0$, denote by $X^{(p)}$ the set of points of codimension $p$ in $X$. For $x \in X^{(p)}$, $\overline{\{x\}}$ is a closed irreducible subscheme of $X$ of codimension $p$. Let $Z^p(X)$ be the free abelian group generated by $X^{(p)}$; hence an element of $Z^p(X)$ can be viewed as a finite integral linear combination of closed irreducible integral subschemes of $X$ of codimension $p$. We call elements of $Z^p(X)$ *p-cycles*. Two $p$-cycles $Z_1, Z_2$ are called *rationally equivalent* if there exist finitely many functions $f_i \in k(y_i)^*$, $y_i \in X^{(p-1)}$, $Y_i := \overline{\{y_i\}}$ such that

$$Z_2 = Z_1 + \sum_i \operatorname{div} f_i,$$

where

$$\operatorname{div} f_i = \sum_{x \in X^{(p)} \cap Y_i} \operatorname{ord}_{\mathcal{O}_{Y_i,x}}(f_i) \cdot \overline{\{x\}}.$$

**Definition 3**  *The p-th Chow group $CH^p(X)$ of $X$ is the quotient group*
$$CH^p(X) := Z^p(X)/\text{rational equivalence} .$$

**Definition 4**  *Two cycles $Y \in Z^p(X), Z \in Z^q(X)$ intersect properly, if $Y \cap Z$ is empty or*
$$\operatorname{codim}_X(Y \cap Z) = \operatorname{codim}_X Y + \operatorname{codim}_X Z \quad (= p + q).$$

Assume $Y \in Z^p(X), Z \in Z^q(X)$ intersect properly. Then, following Serre [Se2], the intersection multiplicity $\chi^x(Y, Z)$ for $x \in Y \cap Z \cap X^{(p+q)}$ is the integer:

$$\chi^x(Y, Z) = \sum_{i \geq 0} (-1)^i \ell_{\mathcal{O}_{X,x}}(\operatorname{Tor}_i^{\mathcal{O}_{X,x}}(\mathcal{O}_{Y,x}, \mathcal{O}_{Z,x})).$$

Recall that, if $R$ is a ring and $M, N$ are $R$-modules; $\operatorname{Tor}_i^R(M, N)$ is defined as follows. Take projective resolutions $P_{\cdot} \to M$ and $Q_{\cdot} \to N$ of $M$ and $N$ respectively. Then $\operatorname{Tor}_i^R(M, N) = H_i(P_{\cdot} \otimes Q_{\cdot})$.

**Theorem 1**  *Assume $X$ is a regular scheme. Then there exists a unique pairing*
$$CH^p(X)_{\mathbb{Q}} \otimes CH^q(X)_{\mathbb{Q}} \longrightarrow CH^{p+q}(X)_{\mathbb{Q}}$$

such that for $Y, Z$ intersecting properly $(Y \in Z^p(X), Z \in Z^q(X))$, we have

$$[Y] \cdot [Z] = [\sum_{x \in Y \cap Z \cap X^{(p+q)}} \chi^x(Y, Z) \cdot \overline{\{x\}}].$$

This theorem is a special case of Theorem 2 below.

For a closed subscheme $Y \subseteq X$ we define $Z_Y^p(X)$ as the group of cycles of codimension $p$ on $X$ supported in the closed subset attached to $Y$. We then define $CH_Y^p(X)$ as the quotient of $Z_Y^p(X)$ by the subgroup generated by the elements $\mathrm{div} f$, where $f \in k(y)^*$ and $y \in X^{(p-1)} \cap Y$.

This group $CH_Y^p(X)$ is the Chow group of codimension $p$ of $X$ with supports in $Y$. Given an inclusion $Y' \subset Y$ of closed subschemes of $X$ there is an obvious map of change of supports

$$CH_{Y'}^p(X) \longrightarrow CH_Y^p(X).$$

**Theorem 2**  *Assume $X$ is a regular scheme and let $Y, Z$ be closed subschemes. Then there exists a pairing*

$$CH_Y^p(X)_{\mathbb{Q}} \otimes CH_Z^q(X)_{\mathbb{Q}} \longrightarrow CH_{Y \cap Z}^{p+q}(X)_{\mathbb{Q}}$$

*satisfying the following properties:*

(i)  $\bigoplus_{Y,p} CH_Y^p(X)_{\mathbb{Q}}$ *is a commutative ring with unit element* $[X] \in CH^0(X)$.

(ii)  *It is compatible with change of supports associated to inclusions* $Y' \subset Y, Z' \subset Z$.

(iii)  *For* $[Y_1] \in CH_Y^p(X), [Z_1] \in CH_Z^q(X)$ *with* $Y_1, Z_1$ *intersecting properly, we have*

$$[Y_1] \cdot [Z_1] = [\sum_{x \in Y_1 \cap Z_1 \cap X^{(p+q)}} \chi^x(Y_1, Z_1) \cdot \overline{\{x\}}].$$

The proof of Theorem 2 will follow from Theorem 3 below; see the discussion after Corollary 2.

## 3. K-Theory

**3.1**    By $K_0(X)$, resp. $K_0'(X)$, we denote the Grothendieck group of coherent locally free, resp. coherent, $\mathcal{O}_X$-modules. For any closed subscheme $Y \subseteq X$, we denote by $K_0^Y(X)$ the Grothendieck group generated by finite complexes

$$\mathcal{F}_{\cdot} : 0 \to \mathcal{F}_n \to \ldots \to \mathcal{F}_1 \to \mathcal{F}_0 \to 0$$

of locally free sheaves on $X$, acyclic outside $Y$, modulo the following two relations:

(i)   $[\mathcal{F}_{\cdot}] = [\mathcal{F}'_{\cdot}]$, if there exists a quasi-isomorphism from $\mathcal{F}_{\cdot}$ to $\mathcal{F}'_{\cdot}$.

(ii)  $[\mathcal{F}_{\cdot}] = [\mathcal{F}'_{\cdot}] + [\mathcal{F}''_{\cdot}]$, if there is an exact sequence

$$0 \to \mathcal{F}'_{\cdot} \to \mathcal{F}_{\cdot} \to \mathcal{F}''_{\cdot} \to 0.$$

From SGA VI and [H], III.6, Exercise 6.9, we have:

**Lemma 1**   *If $X$ is regular, then $K_0(X) \cong K'_0(X)$.*

*Sketch of proof.* One constructs as follows a map $K'_0(X) \to K_0(X)$, which is an inverse of the obvious map $K_0(X) \to K'_0(X)$. Since $X$ is regular, any coherent $\mathcal{O}_X$-module $\mathcal{F}$ has a finite and locally free resolution

$$0 \to \mathcal{F}_n \to \ldots \to \mathcal{F}_1 \to \mathcal{F}_0 \to \mathcal{F} \to 0;$$

see [GBI] or [H], Exercise III.6.9(a). We send $[\mathcal{F}]$ to $\sum_{i=0}^{n}(-1)^i[\mathcal{F}_i] \in K_0(X)$. □

We refer to [GBI] and [H], II.6, Exercise 6.10 for Lemma 2 below. The proof of Lemma 3 is easy.

**Lemma 2**   *If $Y \subseteq X$ is a closed subscheme, there is an exact sequence*

$$K'_0(Y) \to K'_0(X) \to K'_0(X - Y) \to 0.$$

**Lemma 3**   *If $Y, Z \subseteq X$ are closed subschemes, one can define a biadditive pairing*

$$K_0^Y(X) \otimes K_0^Z(X) \to K_0^{Y \cap Z}(X),$$

*by the formula $[\mathcal{F}_{\cdot}] \cdot [\mathcal{G}_{\cdot}] := [\mathcal{F}_{\cdot} \otimes \mathcal{G}_{\cdot}].$*

**Lemma 4**   *If $X$ is regular and $Y \overset{i}{\hookrightarrow} X$ a closed subscheme, then $K_0^Y(X) \cong K'_0(Y)$.*

*Proof.* We first define a map

$$\varphi: K_0^Y(X) \longrightarrow K'_0(Y)$$
$$[\mathcal{F}_{\cdot}] \longmapsto \sum_i (-1)^i [\sum_k \mathcal{J}_Y^k \cdot \mathcal{H}_i(\mathcal{F}_{\cdot})/\mathcal{J}_Y^{k+1} \cdot \mathcal{H}_i(\mathcal{F}_{\cdot})],$$

where $\mathcal{J}_Y$ denotes the ideal sheaf of $Y$; note that the inner sum is actually finite, because there exists $N \in \mathbf{N}$ such that $\mathcal{J}_Y^N$ annihilates $\mathcal{H}_i(\mathcal{F}_{\cdot})$.

We can also define a map in the other direction

$$\psi : K_0'(Y) \longrightarrow K_0^Y(X)$$
$$[\mathcal{E}] \longmapsto [\mathcal{F}_.(\mathcal{E})],$$

where $\mathcal{F}_.(\mathcal{E}) \to i_*\mathcal{E}$ denotes a finite, locally free resolution of $i_*\mathcal{E}$, which exists by the regularity assumption on $X$. One can show that the class of $\mathcal{F}_.(\mathcal{E})$ does not depend on the specific resolution and is additive on exact sequences, hence $\psi$ is well-defined.

One immediately checks $\varphi \circ \psi = \text{id}$. So we are left with showing that $\psi$ is surjective. Given $[\mathcal{F}_.]$, a generator of $K_0^Y(X)$, we have to show that $[\mathcal{F}_.]$ is in the image of $\psi$. We do it by induction on the number of non-zero homology groups of $\mathcal{F}_.$. If $\mathcal{F}_.$ is acyclic, then $[\mathcal{F}_.] = 0$. In general, if $n = \sup\{i|\mathcal{H}_i(\mathcal{F}_.) \neq 0\}$, consider a surjective morphism

$$\sigma : \mathcal{G}_{n+1} \twoheadrightarrow \ker(\partial : \mathcal{F}_n \to \mathcal{F}_{n-1}),$$

with $\mathcal{G}_{n+1}$ locally free. Let

$$\mathcal{F}_{n+1}' = \mathcal{F}_{n+1} \oplus \mathcal{G}_{n+1}$$

be mapped to $\mathcal{F}_n$ by $\partial' = \partial \oplus \sigma$, and let

$$\sigma : \mathcal{G}_{n+2} \twoheadrightarrow \ker(\partial' : \mathcal{F}_{n+1}' \to \mathcal{F}_n)$$

be a surjective morphism with $\mathcal{G}_{n+2}$ locally free. By iterating this construction (which will end when $\partial'$ becomes injective) we get a new complex $\mathcal{F}'$ (equal to $\mathcal{F}_.$ in degree less than $n+1$) and an exact sequence

$$0 \to \mathcal{F}_. \to \mathcal{F}_.' \to \mathcal{G}_. \to 0.$$

Here $\mathcal{F}'$ has one less homology group than $\mathcal{F}_.$, so that, by induction hypothesis, its class is in the image of $\psi$. On the other hand, one checks easily that $\mathcal{G}_.$ is a resolution of $\mathcal{H}_n(\mathcal{F}_.)$, so it is also in the image of $\psi$. We conclude that $[\mathcal{F}_.] = [\mathcal{F}_.'] - [\mathcal{G}_.]$ lies in the image of $\psi$.    □

**3.2**    Let $X, X'$ be noetherian and separated schemes and $f : X \to X'$ a morphism. We get a morphism $f^* : K_0(X') \to K_0(X)$ by pulling back locally free sheaves: $f^*([\mathcal{F}']) = [f^*\mathcal{F}']$.

On the other hand, if $f$ is *proper*, there is a map $f_* : K_0'(X) \to K_0'(X')$ defined by means of the higher direct images:

$$f_*([\mathcal{F}]) = \sum_i (-1)^i [R^i f_* \mathcal{F}];$$

note that for a coherent sheaf $\mathcal{F}$, $R^i f_* \mathcal{F}$ is also coherent, if $f$ is proper, cf. EGA III.3.2.1. Notice the *projection formula*:

$$f_*(f^*[\mathcal{F}] \cdot [\mathcal{G}]) = [\mathcal{F}] \cdot f_*([\mathcal{G}])$$

for $[\mathcal{F}] \in K_0(X')$, $[\mathcal{G}] \in K_0'(X)$ (note $K_0'(X)$ is a $K_0(X)$-module by means of the tensor product); this formula follows from the isomorphisms

$$R^i f_*(f^* \mathcal{F} \otimes \mathcal{G}) = \mathcal{F} \otimes R^i f_* \mathcal{G},$$

cf. [H], III.8, Exercise 8.3.

On $K_0^Y(X)$ we define a decreasing filtration

$$K_0^Y(X) = F^0 K_0^Y(X) \supset F^1 K_0^Y(X) \supset \ldots \supset F^d K_0^Y(X)$$
$$\supset F^{d+1} K_0^Y(X) = \{0\} \quad (d = \dim X),$$

by

$$F^p K_0^Y(X) := \bigcup_{\substack{Z \subseteq Y \\ \mathrm{codim}_X Z \geq p}} \mathrm{im}(K_0^Z(X) \to K_0^Y(X)).$$

Let $Gr^p K_0^Y(X) = F^p K_0^Y(X)/F^{p+1} K_0^Y(X)$.

If $Z \in Z_Y^p(X)$ is an irreducible $p$-cycle contained in $Y$, we can choose a finite, locally free resolution $\mathcal{F}_{\cdot}$ of $i_* \mathcal{O}_Z$, where $i : Z \hookrightarrow X$ is the closed immersion. Denote by $\alpha(Z) \in F^p K_0^Y(X)$ the class of $\mathcal{F}_{\cdot}$.

### 3.3

**Theorem 3** [GS1] *Let $X$ be a finite-dimensional regular scheme.*

(i)   $F^p K_0^Y(X)_{\mathbb{Q}} \cdot F^q K_0^Z(X)_{\mathbb{Q}} \subseteq F^{p+q} K_0^{Y \cap Z}(X)_{\mathbb{Q}}$, *the product being defined as in Lemma 3.3.*

(ii)  *The map $\alpha$ induces an isomorphism*

$$\alpha_{\mathbb{Q}} : CH_Y^p(X)_{\mathbb{Q}} \longrightarrow Gr^p K_0^Y(X)_{\mathbb{Q}}.$$

(iii) *For any morphism $f : X \to X'$ we have, with $Y = f^{-1}(Y')$,*

$$f^* F^p K_0^{Y'}(X')_{\mathbb{Q}} \subset F^p K_0^Y(X)_{\mathbb{Q}}.$$

The proof of this theorem will be given in Section 6.
Before going on, we give two corollaries to Theorem 3.

**Corollary 1** (Serre's vanishing conjecture) *Let $R$ be a regular local ring with maximal ideal $\mathbf{m}$ and residue field $k = R/\mathbf{m}$. Let $M, N$ be finitely generated $R$-modules. Put $X = \mathrm{Spec} R$, $Y = \mathrm{Supp} M$, $Z = \mathrm{Supp} N$ and suppose $Y \cap Z = \mathbf{m}$. If $\mathrm{codim}_X Y + \mathrm{codim}_X Z > \dim X$, then*

$$\chi^R(M, N) := \sum_i (-1)^i \ell_R(\mathrm{Tor}_i^R(M, N)) = 0.$$

*Proof.* By Lemma 4, we have isomorphisms

$$\varphi : K_0^{Y \cap Z}(X) = K_0^{\{\mathbf{m}\}}(\mathrm{Spec} R) \cong K_0'(k) \cong \mathbb{Z},$$

where the last isomorphism is given by the dimension.

Now let $P. \to M, Q. \to N$ be projective resolutions of $M, N$ and denote by $[P.], [Q.]$ their classes in $K_0^Y(X), K_0^Z(X)$ respectively. For the product $[P.] \cdot [Q.]$ we get by the very definition

$$\varphi([P.] \cdot [Q.]) = \sum_i (-1)^i \sum_j \dim_k(\mathbf{m}^j H_i(P. \otimes Q.)/\mathbf{m}^{j+1})$$

$$= \sum_i (-1)^i \ell_R(\mathrm{Tor}_i^R(M, N)) = \chi^R(M, N).$$

On the other hand, by Theorem 3, $[P.] \cdot [Q.] \in F^{p+q}K_0^{Y \cap Z}(X)$, which vanishes, because of the assumption $\mathrm{codim}_X Y + \mathrm{codim}_X Z > \dim X$. Hence

$$\chi^R(M, N) = 0.$$

$\square$

It is still an open problem whether $\chi^R(M, N)$ is always greater or equal to zero as conjectured by Serre [Se 2].

**Corollary 2**  *Let $X$ be a regular scheme of dimension $d$, $E$ a vector bundle and $L$ a line bundle on $X$. Then $[E \otimes L^n] \in K_0(X)_{\mathbb{Q}}$ is a polynomial of degree $d$ with respect to $n$, the coefficients being certain classes of vector bundles.*

*Proof.*    Let $U$ be an open set in $X$ such that $L|_U \cong \mathcal{O}_U$; put $Z := X \backslash U$. Now the class $\alpha := [L] - [\mathcal{O}_X]$ lies in $\mathrm{Ker}(K_0(X) \to K_0(U)) = \mathrm{im}(K_0'(Z) \to K_0(X))$, by Lemma 2. Hence

$$\alpha \in F^1 K_0^X(X) = \bigcup_{Z \subset X} \mathrm{im}(K_0^Z(X) \to K_0^X(X)),$$

using Lemma 1 and Lemma 4. By Theorem 3 we get

$$\alpha^{d+1} \in F^{d+1} K_0^X(X)_{\mathbb{Q}} = 0.$$

If we now expand, we get the desired result:

$$[E \otimes L^n] = [E](\alpha + [\mathcal{O}_X])^n = [E](\alpha + 1)^n$$

$$= \sum_{i=0}^n [E]\alpha^i \binom{n}{i} = \sum_{i=0}^d [E]\alpha^i \binom{n}{i}.$$

$\square$

Note finally that Theorem 3 implies Theorem 2, namely we take the commutativity of the following diagram as the definition of our intersection pairing:

$$CH_Y^p(X)_{\mathbb{Q}} \otimes CH_Z^q(X)_{\mathbb{Q}} \quad \xrightarrow{\ \cong\ } \quad Gr^p K_0^Y(X)_{\mathbb{Q}} \otimes Gr^q K_0^Z(X)_{\mathbb{Q}}$$
$$\downarrow \qquad\qquad\qquad\qquad\qquad\qquad \downarrow$$
$$CH_{Y \cap Z}^{p+q}(X)_{\mathbb{Q}} \quad \xrightarrow{\ \cong\ } \quad Gr^{p+q} K_0^{Y \cap Z}(X)_{\mathbb{Q}}$$

## 4. $\lambda$-rings

### 4.1

**Definition 5**   *A $\lambda$-ring is a unitary ring $R$ with operations $\lambda^k$   $\forall k \geq 0$, satisfying*

(i)   $\lambda^0 = 1, \lambda^1(x) = x$   $\forall x \in R$,   $\lambda^k(1) = 0$   $\forall k > 1$.

(ii)   $\lambda^k(x+y) = \sum_{i=0}^k \lambda^i(x) \cdot \lambda^{k-i}(y)$.

(iii)   $\lambda^k(xy) = P_k(\lambda^1(x), \ldots, \lambda^k(x); \lambda^1(y), \ldots, \lambda^k(y))$ *for some universal polynomials $P_k$ with integer coefficients (defined below).*

(iv)   $\lambda^k(\lambda^\ell(x)) = P_{k,\ell}(\lambda^1(x), \ldots, \lambda^{k\ell}(x))$ *for some universal polynomials $P_{k,\ell}$ with integer coefficients (defined below).*

What we call a $\lambda$-ring is sometimes called a "special $\lambda$-ring".
Putting $\lambda_t(x) := \sum_k \lambda^k(x) t^k$, we have by (ii)

(2)
$$\lambda_t(x+y) = \lambda_t(x) \cdot \lambda_t(y).$$

**4.2**   Let us define $P_k$. We first guess a formula for $P_k$ in the case $x = x_1 + \ldots + x_m$ where $\lambda^k(x_i) = 0$   $\forall k > 1, i = 1, \ldots, m$ and $y = y_1 + \ldots + y_n$ where $\lambda^k(y_j) = 0$   $\forall k > 1, j = 1, \ldots, n$. By (2) we find

$$\lambda_t(x) = \prod_{i=1}^m (1 + t x_i),$$

hence

$$\lambda^k(x) = \begin{cases} \sigma_k(x_1, \ldots, x_m) & k = 1, \ldots, m \\ 0 & k > m, \end{cases}$$

$\sigma_k$ being the elementary symmetric function of degree $k$. Let us impose the following condition on $\lambda^k$: if $\lambda^k(x) = \lambda^k(y) = 0$ for $k > 1$ then $\lambda^k(xy) = 0$ for $k > 1$. Then we get

$$\lambda_t(xy) = \prod_{i=1}^m \prod_{j=1}^n (1 + t x_i y_j).$$

Hence $\lambda^k(xy)$ is a symmetric polynomial in $x_1, \ldots, x_m$ of degree $k$ and

$y_1, \ldots, y_n$ of degree $k$, and so it can be expressed as a polynomial in the elementary symmetric polynomials up to degree $k$:

$$\lambda^k(xy) = P_k(\sigma_1(x_i), \ldots, \sigma_k(x_i); \sigma_1(y_j), \ldots, \sigma_k(y_j))$$
$$= P_k(\lambda^1(x), \ldots, \lambda^k(x); \lambda^1(y), \ldots, \lambda^k(y)).$$

This is our definition of $P_k$.

**4.3**    Similarly, we shall define $P_{k,\ell}$ by first guessing a formula for $P_{k,\ell}$ in the case $x = x_1 + \ldots + x_m$ where $\lambda^k(x_i) = 0$ for all $k > 1$ and then taking this formula as a definition. By 4.2 we have

$$\lambda^\ell(x) = \sum_{1 \le i_1 < \ldots < i_\ell \le m} x_{i_1} \cdot \ldots \cdot x_{i_\ell},$$

hence $\lambda^k(\lambda^\ell(x))$ is the coefficient of $t^k$ in

$$\prod_{1 \le i_1 < \ldots < i_\ell \le m} (1 + t \cdot x_{i_1} \cdot \ldots \cdot x_{i_\ell}).$$

But this is a symmetric polynomial in $x_1, \ldots, x_m$ of degree $k\ell$, hence it can be expressed as a polynomial in the elementary symmetric polynomials up to degree $k\ell$:

$$\lambda^k(\lambda^\ell(x)) = P_{k,\ell}(\sigma_1(x_i), \ldots, \sigma_{k\ell}(x_i))$$
$$= P_{k,\ell}(\lambda^1(x), \ldots, \lambda^{k\ell}(x)).$$

**4.4**    In order to check universal relations among operations on $\lambda$-rings (i.e. polynomials in the $\lambda^k$), it is sufficient to check these on elements of the form $x = x_1 + \ldots + x_N$ with $\lambda^k(x_i) = 0$ for all $k > 1$ $(i = 1, \ldots, N)$ (this is the 'Verification Principle', cf. [AT], §I.3). Furthermore, there is the following 'Splitting Principle' (cf. [AT], §I.6): if $R$ is a $\lambda$-ring and $x \in R$ with $\lambda^k(x) = 0 \quad \forall k > N$, then there exists a $\lambda$-ring $R' \supset R$ such that $x = x_1 + \ldots + x_N$ with $x_i \in R'$ and $\lambda^k(x_i) = 0$ (for all $k > 1$ and $i = 1, \ldots, N$).

**4.5**    Put

$$\psi_{-t}(x) := -t \cdot \frac{d\lambda_t(x)/dt}{\lambda_t(x)}$$

and

$$\psi_t(x) := \sum_{k \ge 1} \psi^k(x) t^k.$$

**Definition 6**    $\psi^k$ *are called the Adams operations on the $\lambda$-ring $R$.*

From the remark in 4.1 we see that the Adams operations are additive:
$$\psi^k(x+y) = \psi^k(x) + \psi^k(y).$$
Now let $x, y \in R$ with $\lambda^k(x) = \lambda^k(y) = 0$ for all $k > 1$ (hence $\lambda^k(xy) = 0$ for all $k > 1$). Then $\lambda_t(xy) = 1 + txy$ and therefore
$$\psi_{-t}(xy) = -t\frac{xy}{1+txy},$$
$$\psi_t(xy) = \frac{txy}{1-txy} = \sum_{k \geq 1}(xy)^k t^k$$
$$\psi^k(xy) = (xy)^k = x^k y^k = \psi^k(x)\psi^k(y).$$
By the verification principle and the additivity of $\psi^k$, this is true for all $x, y \in R$, i.e. the Adams operations constitute ring endomorphisms.

With the verification principle one can also check, for any $x \in R$,
$$\psi^k(\psi^\ell(x)) = \psi^{k\ell}(x).$$

The Adams operations can be expressed in terms of the $\lambda$-operations. To check this, by the verification principle, we may assume that $x = x_1 + \ldots + x_m$, where $\lambda^k(x_i) = 0$ for all $k > 1$ and $i = 1, \ldots, m$. In that case
$$\psi^k(x) = \psi^k(x_1 + \ldots + x_m) = \sum_{i=1}^m \psi^k(x_i)$$
$$= \sum_{i=1}^m x_i^k = N_k(\sigma_1(x_i), \ldots, \sigma_k(x_i))$$
$$= N_k(\lambda^1(x), \ldots, \lambda^k(x)),$$
where $N_k$ denotes the $k$-th Newton polynomial, which expresses $\sum_{i=1}^m x_i^k$ in terms of the elementary symmetric functions of degree less or equal to $k$.

**4.6**

**Theorem 4** *Let $X$ be a regular scheme. Then there exists a $\lambda$-ring structure on $\bigoplus_{Y \subseteq X} K_0^Y(X)$ with the following properties:*

(i) $\lambda^k$ *maps $K_0^Y(X)$ to itself for every $k \geq 0$.*
(ii) *The $\lambda$-ring structure is functorial.*
(iii) *If $X = \mathrm{Spec}R, Y = \mathrm{Spec}(R/aR)$ $(a \in R)$, then for the class of the Koszul complex $\mathrm{Kos}(a)$ in $K_0^Y(X)$, we have*
$$\psi^k([\mathrm{Kos}(a)]) = k \cdot [\mathrm{Kos}(a)].$$

Here $\mathrm{Kos}(a)$ is the complex defined by $0 \to R \xrightarrow{a} R \to 0$, with $R$ in degrees 0 and 1.

The proof of this Theorem will be given in the next section.

## 5. Proof of Theorem 4

**5.1**    We first show that $K_0(X)$ is a $\lambda$-ring. In order to define a $\lambda$-ring structure on $K_0(X)$, one is led to take the $k$-th exterior powers for the operations $\lambda^k$. The problem now consists in verifying the properties (iii), (iv) of Definition 5, (i) and (ii) being obviously true. We solve this problem by constructing an action of the Grothendieck group of certain representations, known to be a $\lambda$-ring, on $K_0(X)$.

Let us be more precise. Let $M_n$ denote the set of $n \times n$ matrices and $H = \mathbf{Z}[M_n] = \mathbf{Z}[X_{11}, X_{12}, \ldots, X_{nn}]$ the associated Hopf algebra over $\mathbf{Z}$, the coproduct $\mu \colon H \to H \otimes_{\mathbf{Z}} H$ being defined as

$$\mu(X_{ij}) := \sum_{k=1}^{n} X_{ik} \otimes X_{kj}.$$

Denote by $R_{\mathbf{Z}}(M_n)$ the Grothendieck group of (isomorphism classes of) $H$-left comodules, free and finitely generated over $\mathbf{Z}$. The homomorphism

$$
(3) \qquad \begin{array}{rcl}
\mathrm{id}_n : \mathbf{Z}^n & \longrightarrow & H \otimes_{\mathbf{Z}} \mathbf{Z}^n \\
e_i & \mapsto & \sum_{j=1}^{n} X_{ij} \otimes e_j,
\end{array}
$$

where $e_1, \ldots, e_n$ is the obvious basis of $\mathbf{Z}^n$, gives $\mathbf{Z}^n$ the structure of an $H$-left comodule. In other words, the diagram

$$
\begin{array}{ccc}
\mathbf{Z}^n & \xrightarrow{\mathrm{id}_n} & H \otimes_{\mathbf{Z}} \mathbf{Z}^n \\
\downarrow{\mathrm{id}_n} & & \downarrow{1 \otimes \mathrm{id}_n} \\
H \otimes_{\mathbf{Z}} \mathbf{Z}^n & \xrightarrow{\mu \otimes 1} & H \otimes_{\mathbf{Z}} H \otimes_{\mathbf{Z}} \mathbf{Z}^n
\end{array}
$$

commutes. We denote the corresponding element in $R_{\mathbf{Z}}(M_n)$ by $\lambda^1(\mathrm{id}_n)$. When $k > 1$ we then denote by $\lambda^k(\mathrm{id}_n)$ the element of $R_{\mathbf{Z}}(M_n)$ obtained by taking the $k$-th exterior power of the representation $\mathrm{id}_n$, defined by (3).

**Theorem 5**    [Se1], [Kr]. *The ring $R_{\mathbf{Z}}(M_n)$ is isomorphic to the polynomial ring*

$$\mathbf{Z}[\lambda^1(\mathrm{id}_n), \ldots, \lambda^n(\mathrm{id}_n)].$$

*The exterior powers define a $\lambda$-ring structure on $R_{\mathbf{Z}}(M_n)$.*

*Proof.* We will just give a brief description and refer the reader to [Se1] and [Kr] for details. By the theory of characters, one knows that the Grothendieck group of representations over $\mathbf{Q}$ of the linear group is

$$R_{\mathbf{Q}}(GL_n) = \mathbf{Z}[\lambda^1(\mathrm{id}_n), \ldots, \lambda^n(\mathrm{id}_n), \lambda^n(\mathrm{id}_n)^{-1}].$$

This implies that

$$R_{\mathbf{Q}}(M_n) = \mathbf{Z}[\lambda^1(\mathrm{id}_n), \ldots, \lambda^n(\mathrm{id}_n)].$$

Similarly, for any finite field $\mathbf{F}_p$, one has

$$R_{\mathbf{F}_p}(M_n) = \mathbf{Z}[\lambda^1(\mathrm{id}_n), \ldots, \lambda^n(\mathrm{id}_n)].$$

As shown in [Se1] §2.4, one has an exact sequence

$$\coprod_{p \text{ prime}} R_{\mathbf{F}_p}(M_n) \xrightarrow{j} R_{\mathbf{Z}}(M_n) \xrightarrow{i} R_{\mathbf{Q}}(M_n) \longrightarrow 0,$$

where $i$ is defined by extension of scalars and $j$ is given as follows. For $[M] \in R_{\mathbf{F}_p}(M_n)$, i.e. $M$ an $\mathbf{F}_p[M_n]$-left comodule, finitely generated over $\mathbf{F}_p$, there is an exact sequence $0 \to P_1 \to P_0 \to M \to 0$ with $P_0, P_1$ being $H$-left comodules, free and finitely generated over $\mathbf{Z}$. The element $[P_0] - [P_1]$ in $R_{\mathbf{Z}}(M_n)$ does not depend on the choice of this exact sequence, cf. [Se1], §2.3. We put $j_p([M]) := [P_0] - [P_1]$ and $j = (j_p)$. On the other hand, by reduction modulo $p$, one gets a homomorphism $d_p : R_{\mathbf{Z}}(M_n) \to R_{\mathbf{F}_p}(M_n)$, and, as in [Se1], §2.7, one finds $j_p d_p = 0$, hence $j = 0$, because $d_p$ is surjective (each $\lambda^k(\mathrm{id}_n)$ lifts). From this the first assertion follows.

To prove the second assertion, we note from the previous result that $R_{\mathbf{Z}}(GL_n) \cong R_{\mathbf{Q}}(GL_n)$ injects into $R_{\mathbf{Q}}(T_n)$, the Grothendieck group of representations over $\mathbf{Q}$ of the group of diagonal matrices of size $n$. But one can check easily that exterior powers induce a $\lambda$-ring on the latter group, since all representations of $T_n$ are direct sums of one-dimensional representations, cf. [Se1], §3.6. $\qquad\square$

**5.2** Let $A$ be a unitary ring. Any $m = (m_{ij}) \in M_n(A)$ defines a homomorphism $m : H = \mathbf{Z}[M_n] \to A$ by evaluating polynomials at $m_{ij}$. Hence any representation $\rho : V \to H \otimes_{\mathbf{Z}} V$ ($V$ a left $H$-comodule, free and finitely generated over $\mathbf{Z}$) induces an action $\rho(m)$ on $A \otimes_{\mathbf{Z}} V$, namely the composition

$$A \otimes_{\mathbf{Z}} V \xrightarrow{\mathrm{id} \otimes \rho} A \otimes_{\mathbf{Z}} (H \otimes_{\mathbf{Z}} V) \xrightarrow{m \otimes \mathrm{id}} A \otimes_{\mathbf{Z}} V.$$

Let now $P$ be a projective $A$-module, i.e. a direct factor of $A^n$ for some $n \in \mathbf{N}$: there exists an $A$-module $Q$ with $A^n = P \oplus Q$. Denote by $p \in M_n(A)$ the projector of $A^n$ onto $P$. For any representation $\rho_n$ of $M_n$ define

$$\rho_n([P]) = [\text{im } \rho_n(p)] \quad \text{in } K_0(A).$$

This element is well-defined since replacing $Q$ by $Q'$ leads to replace $p$ by $p' = g^{-1}p\,g$ for some $g \in GL_n(A)$, hence does not affect the class $[\text{im}\,\rho_n(p)] \in K_0(A)$ of the image of the projector $\rho_n(p)$. Furthermore, given an exact sequence of representations

$$0 \to \rho_n^1 \to \rho_n^2 \to \rho_n^3 \to 0$$

we get an exact sequence of $A$-modules

$$0 \to \text{im } \rho_n^1(p) \to \text{im } \rho_n^2(p) \to \text{im } \rho_n^3(p) \to 0.$$

Finally, this construction is compatible with the standard inclusion $M_n \to M_{n+1}$. So to every $P$ we have attached a ring morphism from $R_{\mathbf{Z}}(M_\infty) = \varprojlim_n R_{\mathbf{Z}}(M_n)$ to $K_0(A)$.

**5.3**    The construction above extends to an arbitrary scheme $X$ as follows. Let $\mathcal{F}$ be a coherent locally free sheaf on $X$. Cover $X$ by affine open subsets $U = \text{Spec}A$. The restriction $\mathcal{F}|_U$ of $\mathcal{F}$ to $U$ is the sheaf attached to a finitely generated projective $A$-module $P$. Using 5.2, given a representation $\rho_n$ of $M_n$, we define $\rho_n(P)$ and these modules can be glued together to give a coherent locally free sheaf $\rho_n(\mathcal{F})$ on $X$. As in the affine case, $\mathcal{F}$ defines a ring morphism from $R_{\mathbf{Z}}(M_\infty)$ to $K_0(X)$.

We are now able to show that $K_0(X)$ is a $\lambda$-ring. Indeed, the image of $(\lambda^k(\text{id}_n)) \in R_{\mathbf{Z}}(M_\infty)$ by the above map is $\lambda^k(\mathcal{F})$, the $k$-th exterior power of $\mathcal{F}$. Applying Theorem 5 we conclude that these satisfy the relations (i)–(iv) of Definition 5 (for more details, cf. [GS1]).

**5.4**    We now proceed to the general situation, i.e. we show that there exists a $\lambda$-ring structure on $\bigoplus_{Y \subseteq X} K_0^Y(X)$ satisfying the properties stated in Theorem 4. For that purpose, we construct operations $\lambda^k : K_0^Y(X) \to K_0^Y(X)$ satisfying (i)–(iv) of Definition 5.

If $\mathcal{F}$ is a finite complex of locally free sheaves on $X$, acyclic outside $Y$, one is of course tempted to define the $\lambda$-ring structure by $\lambda^k(\mathcal{F}.) := (\lambda^k\mathcal{F}.)$, i.e., taking the $k$-th exterior power of each member of the complex $\mathcal{F}.$. But one easily verifies by counting dimensions that $\lambda^k(\mathcal{F}.)$ will no longer be, in general, acyclic outside $Y$. Therefore, we have to consider a different definition. For this, we shall use the notion of simplicial objects in an abelian category and its relation to the category of chain complexes, which we now describe in more detail, following the article of Dold and Puppe [DP].

Let $\Delta$ denote the category of totally ordered finite sets and monotonic maps. So the objects of $\Delta$ are the finite sets $[n] = \{0 < 1 < 2 < \ldots < n\}$ and the morphisms of $\Delta$ are generated by the maps

$$\partial_i : [n-1] \longrightarrow [n], \quad \text{defined by } \partial_i(j) = \begin{cases} j & j < i \\ j+1 & j \geq i \end{cases}$$

$$\sigma_i : [n] \longrightarrow [n-1], \quad \text{defined by } \sigma_i(j) = \begin{cases} j & j \leq i \\ j-1 & j > i. \end{cases}$$

The maps $\partial_i$ (resp. $\sigma_i$) are called faces (resp. degeneracies) for $i = 0, \ldots, n$.

If $\mathcal{C}$ is any category, the category $S\mathcal{C}$ of simplicial objects in $\mathcal{C}$ is defined as follows. The objects of $S\mathcal{C}$ are contravariant functors $S : \Delta \to \mathcal{C}$ and the morphisms are natural transformations between these. Hence, a simplicial object is given by a sequence $\{S_n = S([n])\}$ and, for each monotonic map $\alpha : [m] \to [n]$, a map $S_\alpha := S(\alpha) : S_n \to S_m$, such that

$$S_{\mathrm{id}_{[n]}} = \mathrm{id}_{S_n}$$

and

$$S_{\alpha \circ \beta} = S_\beta \circ S_\alpha.$$

Define $d_i := S_{\partial_i} : S_n \to S_{n-1}, s_i := S_{\sigma_i} : S_{n-1} \to S_n$. A simplicial morphism $f$ between $S$ and $T$ is given by a sequence of maps

$$\{f_n : S_n \to T_n\},$$

such that the following diagram commutes:

$$
\begin{array}{ccc}
S_n & \xrightarrow{f_n} & T_n \\
{\scriptstyle S_\alpha}\downarrow & & \downarrow{\scriptstyle T_\alpha} \\
S_m & \xrightarrow{f_m} & T_m.
\end{array}
$$

**5.5** Now, if $\mathcal{A}$ is an abelian category, i.e. there is a notion of kernel and cokernel, we denote by $C\mathcal{A}$ the category of chain complexes (of homological type) associated to $\mathcal{A}$. We define a functor $N : S\mathcal{A} \to C\mathcal{A}$ as follows. For any object $S$ of $S\mathcal{A}$, put

$$(NS)_n := \bigcap_{i=1}^{n} \ker\left(d_i : S_n \to S_{n-1}\right)$$

for $n \geq 1$ and $(NS)_0 = S_0$. Because $d_i \circ d_j = d_{j-1} \circ d_i$ for $i < j$, $d_0(NS)_n \subset (NS)_{n-1}$ and $d_0^2 = 0$, hence $d_0$ defines a map

$$\partial : (NS)_n \to (NS)_{n-1}$$

with $\partial^2 = 0$, and we define $NS$ to be the chain complex $(NS)_n$ with boundary map $\partial$. If $f : S \to T$ is a simplicial map, we obtain by restriction a chain map $Nf : NS \to NT$.

Conversely, one can also define a functor $K : C\mathcal{A} \to S\mathcal{A}$. For any object $C$ in $C\mathcal{A}$, we put

(4)
$$(KC)_n := \bigoplus_{q \leq n} \bigoplus_{\eta} C_q,$$

where $\eta$ runs over all surjective monotonic maps $\eta : [n] \twoheadrightarrow [q]$. Furthermore, for each monotonic map $\alpha : [m] \to [n]$, we define

$$(KC)_\alpha : (KC)_n \to (KC)_m$$

by giving the restrictions of $(KC)_\alpha$ to each summand in (4) as follows; note that for $\alpha : [m] \to [n]$, $\eta : [n] \twoheadrightarrow [q]$, there exist unique monotonic maps $\eta' : [m] \twoheadrightarrow [p]$, $\epsilon : [p] \hookrightarrow [q]$ for some $p$, such that $\eta \circ \alpha = \epsilon \circ \eta'$:

(a)   If $p = q$, then $(KC)_\alpha|_{C_q} : C_q \to C_q$ is given by the identity map on $C_q$;

(b)   If $p = q - 1$ and $\epsilon(0) = 1$, then $(KC)_\alpha|_{C_q} : C_q \to C_p$ is given by the boundary map $\partial$ on $C_q$;

(c)   In all other cases, $(KC)_\alpha|_{C_q} : C_q \to C_p$ is given by the zero map on $C_q$.

If $f : C \to D$ is a chain map, one checks that it induces a simplicial map $Kf$ between $KC$ and $KD$.

The main result now states (cf. [DP], Satz 3.6) that the functor $NK : C\mathcal{A} \to C\mathcal{A}$ is the identity and that $KN : S\mathcal{A} \to S\mathcal{A}$ is naturally equivalent to the identity functor, i.e., there exist natural transformations $F : KN \to \mathrm{id}_{S\mathcal{A}}$ and $G : \mathrm{id}_{S\mathcal{A}} \to KN$ such that $F \circ G = G \circ F = \mathrm{id}$. As a consequence, $N$ and $K$ are exact functors.

**5.6**    If $G$ is a free abelian group with generators $g_i$ $(i \in I)$, one defines for any $A \in \mathrm{ob}(\mathcal{A})$ the tensor product $G \otimes A \in \mathrm{ob}(\mathcal{A})$ by

$$G \otimes A := \bigoplus_{i \in I} A.$$

This definition can be made independent of the choice of the set of generators (cf. [DP], §3.32).

Using this remark, we can define chain (resp. simplicial) homotopy in $C\mathcal{A}$ (resp. $S\mathcal{A}$). Let $N(1) \in \mathrm{ob}(C\mathcal{A})$ denote the normalized chain complex of the standard 1-simplex, namely $N(1)_n = 0$ for $n > 1$, $N(1)_1 = \mathbf{Z}[e]$, $N(1)_0 = \mathbf{Z}[e_0] \oplus \mathbf{Z}[e_1]$ and $\partial(e) = e_1 - e_0$. A *chain homotopy* between two chain maps $f_0, f_1 : C \longrightarrow D$ in $C\mathcal{A}$ is a morphism

$$h : N(1) \otimes C \longrightarrow D$$

such that $h(e_i \otimes c) = f_i(c)$ $(i = 0, 1)$. The restriction of $h$ to $N(1)_1 \otimes C$ is a map $H : C[-1] \to D$ , and $h$ is a chain map iff $\partial H + H\partial = f_0 - f_1$, i.e. $H$ is a homotopy between the chain maps $f_0$ and $f_1$.

On the other hand, let $K(1) := K(N(1)) \in \mathrm{ob}(S\mathcal{A})$. A *simplicial homotopy* between two morphisms $f_0, f_1 : S \to T$ in $S\mathcal{A}$ is a morphism

$$h : K(1) \otimes S \longrightarrow T$$

such that $h(s_0^n(e_i) \otimes s) = f_i(s)$ $(i = 0, 1)$. Here $e_0, e_1$ denote the two generators of $K(1)_0 = N(1)_0$ and $s_0^n$ is the obvious inclusion

$$K(1)_0 \hookrightarrow K(1)_n = N(1)_0 \oplus \bigoplus_\eta N(1)_1,$$

where $\eta$ runs over all surjective monotonic maps $[n] \twoheadrightarrow [1]$. So, in degree $n$, the map $h_n : S_n \oplus S_n \oplus \bigoplus_\eta S_n \to T_n$ is equal to $f_{0,n}$ (resp. $f_{1,n}$) on the first (resp. second) summand.

By [DP], Satz 3.31, we know that $N$ and $K$ preserve homotopy. In the sequel, we call $C \in \mathrm{ob}(C\mathcal{A})$ (resp. $S \in \mathrm{ob}(S\mathcal{A})$) contractible when the identity map on $C$ (resp. $S$) is homotopic to the zero map.

Let now $F : \mathcal{A} \to \mathcal{B}$ be a functor between the abelian categories $\mathcal{A}, \mathcal{B}$ satisfying $F(0) = 0$. We extend $F$ to a functor $F : C\mathcal{A} \to C\mathcal{B}$ by setting

$$F(C) := NFK(C).$$

According to [DP], Satz 3.31, this functor will preserve homotopy. Indeed, we just need to check it for simplicial objects and, given a homotopy $h : K(1) \otimes S \to T$, we get the homotopy

$$F(h) : K(1) \otimes FS \longrightarrow FT,$$

by composing $h$ with the natural map

$$K(1) \otimes FS \longrightarrow F(K(1) \otimes S).$$

**5.7** We are now able to define operations $\lambda^k : K_0^Y(X) \to K_0^Y(X)$. Let $\mathcal{P}(X)$ be the category of locally free coherent sheaves on $X$ and $F = \lambda^k : \mathcal{P}(X) \to \mathcal{P}(X)$ be the $k$-th exterior power. By the above procedure, we define

$$\lambda^k(\mathcal{F}.) := N\lambda^k K(\mathcal{F}.);$$

here $\mathcal{P}(X)$ is embedded in the abelian category of all coherent $\mathcal{O}_X$-modules. One checks that $\lambda^k(\mathcal{F}_\cdot)$ is also in $C\mathcal{P}(X)$. Therefore $\lambda^k$ induces a homotopy preserving functor from $C\mathcal{P}(X)$ into itself (see §5.5). If $\mathcal{F}_\cdot$ is acyclic outside $Y$, the restriction to any open affine set in $X\backslash Y$ is also acyclic and therefore contractible; hence $\lambda^k(\mathcal{F}_\cdot)$ is also acyclic outside $Y$. Similarly, given any quasi-isomorphism from $\mathcal{F}_\cdot$ to $\mathcal{G}_\cdot$, the induced morphism from $\lambda^k(\mathcal{F}_\cdot)$ to $\lambda^k(\mathcal{G}_\cdot)$ is an homotopy equivalence on each open affine set, hence a quasi-isomorphism. Finally, we know, by [DP] §4.23, that the complex $\lambda^k(\mathcal{F}_\cdot)$ is finite when the complex $\mathcal{F}_\cdot$ is finite (the length of $\lambda^k(\mathcal{F}_\cdot)$ is at most $k$ times the length of $\mathcal{F}_\cdot$). Therefore we get maps $\lambda^k : K_0^Y(X) \to K_0^Y(X)$, and we need to verify (i)–(iv) of Definition 5. Obviously (i) is true, because, by convention, $\lambda^0(\mathcal{F}) = \mathcal{O}_X$ for any locally free sheaf $\mathcal{F}$. To prove (ii), we first note that, if

$$0 \to \mathcal{F}' \to \mathcal{F} \to \mathcal{F}'' \to 0$$

is an exact sequence in $\mathcal{P}(X)$, then $\lambda^k(\mathcal{F})$ admits a canonical filtration whose successive quotients are canonically isomorphic to $\lambda^{k'}(\mathcal{F}') \otimes \lambda^{k''}(\mathcal{F}'')$ with $k' + k'' = k$. Let now $[\mathcal{F}_\cdot] = [\mathcal{F}'_\cdot] + [\mathcal{F}''_\cdot]$ in $K_0^Y(X)$, i.e. there is an exact sequence in $C\mathcal{P}(X)$

$$0 \longrightarrow \mathcal{F}'_\cdot \longrightarrow \mathcal{F}_\cdot \longrightarrow \mathcal{F}''_\cdot \longrightarrow 0 \ .$$

Since $K$ is exact, $\lambda^k K(\mathcal{F}_\cdot)$ admits a filtration whose successive quotients are isomorphic to $\lambda^{k'} K(\mathcal{F}'_\cdot) \otimes \lambda^{k''} K(\mathcal{F}''_\cdot)$ with $k'+k'' = k$. Since $N$ is also exact, $N\lambda^k K(\mathcal{F}_\cdot) = \lambda^k(\mathcal{F}_\cdot)$ has a filtration whose successive quotients are isomorphic to $N(\lambda^{k'} K(\mathcal{F}'_\cdot) \otimes \lambda^{k''} K(\mathcal{F}''_\cdot))$, which is isomorphic to $N\lambda^{k'} K(\mathcal{F}'_\cdot) \otimes N\lambda^{k''} K(\mathcal{F}''_\cdot)$. Therefore, in the group $K_0^Y(X)$, we get the equality

$$[\lambda^k(\mathcal{F}_\cdot)] = \bigoplus_{i=0}^{k} [\lambda^i(\mathcal{F}'_\cdot)] \otimes [\lambda^{k-i}(\mathcal{F}''_\cdot)],$$

which proves (ii).

To prove (iii) and (iv) we proceed as in the case $Y = X$ considered above. Let $[\rho_\infty] \in R_{\mathbf{Z}}(M_\infty)$ an element of degree $d$ as a polynomial in $\varprojlim_{n}[\lambda^1(\mathrm{id}_n), \dots, \lambda^n(\mathrm{id}_n)]$ and $[\mathcal{F}_\cdot] \in K_0^Y(X)$, with $\mathcal{F}_\cdot$ of length $N$. We define $[\rho_\infty](\mathcal{F}_\cdot) \in K_0^Y(X)$ as follows. Let $n_0$ be such that all the bundles $K(\mathcal{F}_\cdot)_n$, $n \le Nd + 1$, are locally direct factors of $\mathcal{O}_X^{n_0}$. We may then define, as in 5.3, $\rho(K(\mathcal{F}_\cdot)_n)$ for any representation $\rho$ of $M_{n_0}$. Take

$$\rho(\mathcal{F}_\cdot)_n = \bigcap_{i \ge 1} \ker(d_i \text{ on } \rho(K(\mathcal{F}_\cdot)_n))$$

when $n \le Nd+1$. One can show that $\rho(\mathcal{F}_\cdot)_{Nd+1} = 0$, so it makes sense to take $\rho(\mathcal{F}_\cdot)_n = 0$ when $n \ge Nd+1$. This construction is compatible with exact sequences and stabilization, so we get an element $[\rho_\infty](\mathcal{F})$ for any $[\rho_\infty] \in R_{\mathbf{Z}}(M_\infty)$. When $\rho_\infty = \lambda^k(\mathrm{id})$, we have $[\lambda^k(\mathrm{id})](\mathcal{F}) = [\lambda^k(\mathcal{F})]$. Using the fact that $R_{\mathbf{Z}}(M_\infty)$ is a $\lambda$-ring, we conclude that the properties (i)–(iv) in Definition 5 are satisfied by $\lambda^k$ on $K_0^Y(X)$ (see [GS1] for more details).

**5.8** This proves that $\bigoplus\limits_{Y \subseteq X} K_0^Y(X)$ has a $\lambda$-ring structure. By construction this $\lambda$-ring structure is functorial. To complete the proof of Theorem 4, we have to verify that, when $X = \mathrm{Spec}R$ and $Y = \mathrm{Spec}(R/aR)$ ($a \in R$), $\psi^k([\mathrm{Kos}(a)]) = k \cdot [\mathrm{Kos}(a)]$. Here $\mathrm{Kos}(a)$ is the Koszul complex

$$C : 0 \to E \xrightarrow{u} F \to 0,$$

where $E = F = R$ and $u$ is the multiplication by $a$. By the definitions of the functor $K$, we have

$$(KC)_n = F \oplus \bigoplus_{\eta_i} E,$$

where $\eta_i$ are the $n$ surjective monotonic maps from $[n]$ onto $[1]$, given by

$$\eta_i(j) = \begin{cases} 0 & j < i \\ 1 & j \geq i \end{cases} \quad (i = 1, \ldots, n).$$

We write $(KC)_n = F \oplus E_1 \oplus \ldots \oplus E_n$, where $E_i$ is the copy of $E$ belonging to $\eta_i$. Because $\eta_i \circ \partial^j = \eta_i$ for $i \leq j$ and $\eta_i \circ \partial^j = \eta_{i-1}$ for $i > j$, unless $i = 1, j = 0$ or $i = j = n$ (where $\eta_1 \circ \partial^0$ and $\eta_n \circ \partial^n$ are the zero maps), $d_j : (KC)_n \to (KC)_{n-1}$ is given by the formulae

$$d_0(f, e_1, \ldots, e_n) = (f + u(e_1), e_2, \ldots, e_n),$$
$$d_j(f, e_1, \ldots, e_n) = (f, e_1, \ldots, e_{j-1}, e_j + e_{j+1}, \ldots, e_n) \quad \text{if } 0 < j < n,$$
$$d_n(f, e_1, \ldots, e_n) = (f, e_1, \ldots, e_{n-1}).$$

Now $\lambda^k KC$ is given by the sequence

$$(\lambda^k KC)_n = \bigoplus \lambda^\alpha F \otimes \lambda^{\beta_1} E_1 \otimes \ldots \otimes \lambda^{\beta_n} E_n,$$

the sum being taken over all tuples $(\alpha, \beta_1, \ldots, \beta_n) \in \mathbf{N}^{n+1}$ such that $\alpha + \beta_1 + \ldots + \beta_n = k$. The faces $d_j$ on $\lambda^k KC$ are induced by the faces on $KC$ described above. One computes

$$(N\lambda^k KC)_n = \bigcap_{j=1}^n \ker(d_j : (\lambda^k KC)_n \longrightarrow (\lambda^k KC)_{n-1})$$

$$= \bigoplus \lambda^\alpha F \otimes \lambda^{\beta_1} E_1 \otimes \ldots \otimes \lambda^{\beta_n} E_n,$$

the sum being taken over all tuples $(\alpha, \beta_1, \ldots, \beta_n) \in \mathbf{N}^{n+1}$ such that $\alpha + \beta_1 + \ldots + \beta_n = k$ and $\beta_i > 0$. Therefore $\alpha = 0, 1$ and $\beta_i = 1$, since

$F = E_i = R$   $(i = 1, \ldots, n)$. From this we deduce $N\lambda^k KC = C[1 - k]$, which implies

$$\lambda^k [\mathrm{Kos}(a)] = \lambda^k [C] = [N\lambda^k KC]$$
$$= [C[1 - k]] = (-1)^{k-1} [C] \qquad \text{(note that } [C] = -[C[-1]]),$$

and this gives

$$\psi^k ([\mathrm{Kos}(a)]) = k \cdot [\mathrm{Kos}(a)],$$

by noting that $[\mathrm{Kos}(a)]^2 = 0$.

**5.9**

**Remark** If $X$ is defined over $\mathbb{F}_p$, it has a Frobenius endomorphism $F$, which is the identity map on the underlying topological space and raises the sections of the structure sheaf $\mathcal{O}_X$ to the $p$-th power. For any closed subset $Y \subseteq X$ this induces a map $F^* : K_0^Y(X) \to K_0^Y(X)$. One can show that $F^* = \psi^p$. So, in characteristic $p > 0$, Frobenius provides a quick definition of the Adams operation $\psi^p$.

## 6. Proof of Theorem 3

**6.1**   We first prove part (i) of Theorem 3.

**Lemma 5**   *Let $X$ be a regular scheme and $Y \subseteq X$ a closed subscheme with* $\mathrm{codim}_X Y = m$. *Then there exists an exact sequence*

$$0 \to F^{m+1} K_0^Y(X) \to K_0^Y(X) \to \bigoplus_{x \in Y \cap X^{(m)}} K_0^{\{x\}}(\mathcal{O}_{X,x}).$$

*Proof.*   By Lemma 2 we have, for $Z \subseteq Y$ a closed subscheme, the following exact sequence

$$K_0'(Z) \to K_0'(Y) \to K_0'(Y - Z) \to 0.$$

By Lemma 4, it may be written

$$0 \to \mathrm{im}(K_0^Z(X) \to K_0^Y(X)) \to K_0^Y(X) \to K_0^{Y-Z}(X - Z) \to 0.$$

By taking the inductive limit over all closed subschemes $Z \subseteq Y$ with $\mathrm{codim}_X Z \geq m + 1$, we get

$$0 \to F^{m+1} K_0^Y(X) \to K_0^Y(X) \to \varinjlim_{\substack{Z \subseteq Y \\ \mathrm{codim}_X Z \geq m+1}} K_0^{Y-Z}(X - Z) \to 0.$$

The isomorphism

$$\varinjlim_{\substack{Z \subseteq Y \\ \mathrm{codim}_X Z \geq m+1}} K_0^{Y-Z}(X - Z) \cong \bigoplus_{x \in Y \cap X^{(m)}} K_0^{\{x\}}(\mathcal{O}_{X,x})$$

completes the proof.                                                    □

**Definition 7**  *Fix $k > 1$. For $i \geq 0$, we define the* weight $i$-part
*of $K_0^Y(X)_\mathbb{Q}$ to be the subspace $K_0^Y(X)^{(i)}$ made of the elements $\alpha \in$
$K_0^Y(X)_\mathbb{Q}$ such that*

$$\psi^k(\alpha) = k^i \alpha.$$

From Theorem 4.1 it follows that $K_0^Y(X)^{(i)}$ does not depend on the
choice of $k$ (see [S1]).

**Lemma 6**  *With the above notation we have*

$$F^p K_0^Y(X)_\mathbb{Q} = \bigoplus_{i \geq p} K_0^Y(X)^{(i)}.$$

*Proof.* We use descending induction on $m = \mathrm{codim}_X Y$. So suppose first
that $Y = \{x\}$. Then $K_0^{\{x\}}(X) \cong K_0'(\{x\}) \cong K_0(k(x)) \cong \mathbb{Z}$ (the last
isomorphism is given by the dimension, so the only thing to check is that
the action of $\psi^k$ on a single non-zero element of $K_0^{\{x\}}(X) = F^d K_0^Y(X)$
$(d = \dim X)$ is multiplication by $k^d$. For that purpose, let $R$ be the
regular local ring $\mathcal{O}_{X,x}$ with maximal ideal $\mathbf{m} = (a_1, \ldots, a_d)$, $a_1, \ldots, a_d$
denoting a system of parameters. By [Ma], p. 135, we then know that

$$\mathrm{Kos}(a_1, \ldots, a_d) = \bigotimes_{i=1}^d \mathrm{Kos}(a_i)$$

is a resolution of $R/\mathbf{m} = k(x)$ (because $a_1, \cdots, a_d$ is an $R$-regular sequence), so it defines an element in $K_0^{\{x\}}(X)$. We now compute $\psi^k$
of this element using Theorem 4.1 (without loss of generality, we may
replace $X$ by $\mathrm{Spec}\mathcal{O}_{X,x}$):

$$\psi^k([\mathrm{Kos}(a_1, \ldots, a_d)]) = \psi^k(\bigotimes_{i=1}^d [\mathrm{Kos}(a_i)]) = \prod_{i=1}^d k \cdot [\mathrm{Kos}(a_i)]$$
$$= k^d [\mathrm{Kos}(a_1, \ldots, a_d)],$$

as claimed.

We now proceed to the general case, i.e. we assume $\mathrm{codim}_X Y = m <$
$d$, assuming the claim for all closed subschemes of codimension bigger
than $m$.

For $\alpha \in F^p K_0^Y(X)_\mathbb{Q}$, we first show the existence of a decomposition
$\alpha = \sum_{i \geq p} \alpha_i$, where $\psi^k(\alpha_i) = k^i \alpha_i$.

If $p > m$, $\alpha \in F^{m+1} K_0^Y(X)$, i.e. $\alpha \in \mathrm{im}(K_0^Z(X) \rightarrow K_0^Y(X))$ for some

$Z$ with $\text{codim}_X Z \geq m+1$. Hence in this case the decomposition follows by the induction hypothesis.

If $p = m$, look at the exact sequence from Lemma 6.1:

$$0 \to F^{m+1}K_0^Y(X) \to K_0^Y(X) \overset{\epsilon}{\longrightarrow} \bigoplus_{x \in Y \cap X^{(m)}} K_0^{\{x\}}(\mathcal{O}_{X,x}) \to 0.$$

By the above argument $\psi^k(\epsilon(\alpha)) = k^m \epsilon(\alpha)$, hence $\epsilon(\psi^k(\alpha) - k^m \alpha) = 0$. So $\psi^k(\alpha) - k^m \alpha \in F^{m+1}K_0^Y(X)$ and, by the induction hypothesis,

$$\psi^k(\alpha) - k^m \cdot \alpha = \sum_{i>m} \beta_i,$$

where $\psi^k(\beta_i) = k^i \beta_i$  $(i > m)$. Now put

$$\alpha_i := (k^i - k^m)^{-1} \beta_i \qquad ,i > m,$$

$$\alpha_m := \alpha - \sum_{i>m} \alpha_i .$$

Then, if $i > m$, we get

$$(\psi^k - k^i)(\alpha_i) = (k^i - k^m)^{-1}(k^i - k^i)\beta_i = 0.$$

Furthermore,

$$(\psi^k - k^m)(\alpha_m) = \sum_{i>m} \beta_i - \sum_{i>m} (\psi^k - k^m)(\alpha_i)$$

$$= \sum_{i>m} \beta_i - \sum_{i>m} \frac{k^i - k^m}{k^i - k^m} \cdot \beta_i = 0.$$

Hence $\alpha = \sum_{i \geq m} \alpha_i$, with $\psi^k(\alpha_i) = k^i \cdot \alpha_i$, as desired.

The uniqueness of such a decomposition follows also inductively. Assume we had two decompositions

$$\sum_{i \geq m} \alpha_i = \alpha = \sum_{i \geq m} \alpha_i' .$$

Applying $\psi^k$, we derive the two decompositions

$$\sum_{i>m} (k^i - k^m)\alpha_i = \sum_{i>m} (k^i - k^m)\alpha_i' .$$

By the induction hypothesis $\alpha_i = \alpha_i'$ for $i > m$, hence also $\alpha_m = \alpha_m'$, as desired.

So far, we have shown the inclusion

$$F^p K_0^Y(X)_{\mathbb{Q}} \subset \bigoplus_{i \geq p} K_0^Y(X)^{(i)} .$$

The proof of Lemma 6 will be complete if we show the opposite inclusion. So assume $\alpha \in \bigoplus_{i \geq p} K_0^Y(X)^{(i)}$ and $\alpha \in F^q K_0^Y(X) \backslash F^{q+1} K_0^Y(X)$. We have

$$\alpha = \sum_{i \geq p} \alpha_i = \sum_{j \geq q} \beta_j \quad \text{with } \beta_q \neq 0.$$

Hence $\alpha_q \neq 0$, from which $p \leq q$, i.e. $\alpha \in F^p K_0^Y(X)$, follows.   □

We can now easily prove Theorem 3(i). Let $\alpha \in F^p K_0^Y(X)_{\mathbb{Q}}$ and $\beta \in F^q K_0^Z(X)_{\mathbb{Q}}$. By Lemma 6 we then have $\alpha = \sum_{i \geq p} \alpha_i$ with $\alpha_i \in K_0^Y(X)^{(i)}$ and $\beta = \sum_{j \geq q} \beta_j$ with $\beta_j \in K_0^Z(X)^{(j)}$. Therefore, $\alpha\beta = \sum_{i,j} \alpha_i \beta_j$ with

$$\psi^k(\alpha_i \beta_j) = k^{i+j}\alpha_i \beta_j, \quad i + j \geq p + q.$$

So $\alpha\beta$ lies in $F^{p+q} K_0^{Y \cap Z}(X)_{\mathbb{Q}}$, as claimed.

**6.2**   We are left with proving Theorem 3(ii), namely the isomorphism

$$CH_Y^p(X)_{\mathbb{Q}} \cong Gr^p K_0^Y(X)_{\mathbb{Q}}.$$

To achieve this, we have first to review some knowledge of algebraic $K$-theory (cf. [Q1]). Unfortunately, giving more details on these constructions would take us beyond the scope of this book.

**Definition 8**   *A category $\mathcal{E}$ is called* exact *if there exists an abelian category $\mathcal{A}$ such that $\mathcal{E}$ is a full subcategory of $\mathcal{A}$ closed under extension: for any exact sequence $0 \to E' \to E \to E'' \to 0$ in $\mathcal{A}$, if $E'$ and $E''$ are in $\mathrm{ob}(\mathcal{E})$ then $E$ lies in $\mathrm{ob}(\mathcal{E})$.*

For an exact category $\mathcal{E}$, Quillen [Q1] defined $K$-groups $K_m(\mathcal{E})$ ($m \in \mathbb{N}$) by introducing a simplicial set $BQ\mathcal{E}$, constructed out of the objects and exact sequences of $\mathcal{E}$, and putting

$$K_m(\mathcal{E}) := \pi_{m+1}(BQ\mathcal{E}),$$

the $(m + 1)$-th homotopy group of $BQ\mathcal{E}$; this definition does not depend on the choice of the abelian category $\mathcal{A}$. Furthermore $K_0(\mathcal{E})$ is isomorphic to the Grothendieck group of $\mathcal{E}$.

In our situation, $X$ a noetherian, separated scheme and $Y \subseteq X$ a closed subset, we use the following exact categories: $\mathcal{M}(X)$, the category of coherent sheaves on $X$; $\mathcal{M}_Y(X)$, the category of coherent sheaves on $X$ supported on $Y$; $\mathcal{P}(X)$, the category of coherent locally free sheaves on $X$.

For $m \in \mathbb{N}$ we then put

$$K'_m(X) := K_m(\mathcal{M}(X));$$
$$K'^Y_m(X) := K_m(\mathcal{M}_Y(X));$$
$$K_m(X) := K_m(\mathcal{P}(X)).$$

We note that these definitions are compatible with the previous ones in the case $m = 0$, due to the above mentioned isomorphism between $K_0(\mathcal{E})$

and the Grothendieck group of $\mathcal{E}$. Furthermore $K_m'^Y(X)$ is isomorphic to $K_m'(Y)$.

If $A$ is a unitary commutative ring, we write $K_m(A)$ instead of $K_m(\text{Spec } A)$. For example, when $A = F$ is a field, we have (cf. [Mi], §§1, 3, 11):

$$K_0(F) \cong \mathbb{Z},$$
$$K_1(F) \cong F^*,$$
$$K_2(F) \cong F^* \otimes_{\mathbb{Z}} F^* / \langle x \otimes (1-x) : x \in F^* \backslash \{1\} \rangle.$$

In other words, $K_2(F)$ is defined by the generators $\{x, y\}$, $x, y \in F^* \backslash \{1\}$, and the relations

$$\{x_1 x_2, y\} = \{x_1, y\} + \{x_2, y\},$$
$$\{x, y_1 y_2\} = \{x, y_1\} + \{x, y_2\}, \quad \text{and}$$
$$\{x, 1-x\} = 0.$$

The tensor product of modules induces a product

$$K_m(X) \otimes K_n(X) \to K_{m+n}(X),$$

making $\bigoplus_{m \geq 0} K_m(X)$ into a ring. It gives also a product

$$K_m(X) \otimes K_n'(X) \to K_{m+n}'(X),$$

making $\bigoplus_{m \geq 0} K_m'(X)$ into a $\bigoplus_{m \geq 0} K_m(X)$-left module.

The product by the class of $\mathcal{O}_X$ defines a natural map $K_m(X) \to K_m'(X)$. This is an isomorphism if $X$ is regular, cf. Lemma 4 in the case $m = 0$.

## 6.3

**Definition 9** *A subcategory $S$ of an exact category $\mathcal{E}$ is called a Serre subcategory if $S$ is a full subcategory and for any exact sequence $0 \to E' \to E \to E'' \to 0$ in $\mathcal{E}$, $E', E'' \in \text{ob}(S)$ is equivalent to $E \in \text{ob}(S)$.*

In the situation of Definition 9, there exists a quotient category $\mathcal{E}/S$, which is again an exact category. In this context Quillen established the following long exact sequence, called the *localization exact sequence* (cf. [Q1])

$$\ldots \to K_{m+1}(\mathcal{E}/S) \to K_m(S) \to K_m(\mathcal{E}) \to K_m(\mathcal{E}/S) \to K_{m-1}(S) \to \ldots .$$

For example, let $A$ be a Dedekind domain, $F$ its quotient field, $\mathcal{E}$ the exact category of finitely generated $A$-modules, and $S$ the Serre subcategory of torsion $A$-modules. Observing $\mathcal{E}/S \cong \mathcal{P}(\text{Spec} F)$ and $S \cong \coprod_{\wp \neq 0} \mathcal{P}(\text{Spec} k(\wp))$, where $\wp$ runs over all non-zero prime ideals of

$A$ and $k(\wp)$ denotes the residue field $A/\wp$, we obtain the exact sequence

$$\ldots \to K_{m+1}(F) \to \bigoplus_{\wp \neq 0} K_m(k(\wp)) \to K_m(A) \to K_m(F)$$

$$\to \bigoplus_{\wp \neq 0} K_{m-1}(k(\wp)) \to \ldots \; .$$

In particular, the map

$$F^* \cong K_1(F) \longrightarrow \bigoplus_{\wp \neq 0} K_0(k(\wp)) \cong \bigoplus_{\wp \neq 0} \mathbf{Z}$$

is nothing but the valuation map $f \mapsto (v_\wp(f))$, $f \in F^*$.

The localization exact sequence leads also to the following. Let $X, Y$ be as above and $\dim X < \infty$. Denote by $\mathcal{M}_Y^p(X)$ the category of coherent sheaves $\mathcal{F}$ on $X$ supported on $Y$ and such that $\mathrm{codim}_X(\mathrm{supp}\mathcal{F}) \geq p$. These give a filtration of the exact category $\mathcal{M}_Y(X)$ by Serre subcategories.

**Theorem 6**   *There exists a spectral sequence $E_{rY}^{p,q}(X)$ with differentials*
$d_r^{p,q} : E_{rY}^{p,q}(X) \longrightarrow E_{rY}^{p+r,q-r+1}(X)$ *with*

$$E_{1Y}^{p,q}(X) = \begin{cases} K_{-p-q}(\mathcal{M}_Y^p(X)/\mathcal{M}_Y^{p+1}(X)) & p \geq 0, p+q \leq 0 \\ 0 & \text{otherwise} \end{cases}$$

*and converging to $K_{-p-q}'^Y(X)$.*

*Proof.*   The long exact sequence for $\mathcal{E} = \mathcal{M}_Y^p(X)$ and $\mathcal{S} = \mathcal{M}_Y^{p+1}(X)$ leads to the exact couple

$$\bigoplus_{\substack{m \geq 0 \\ p \geq 0}} K_m(\mathcal{M}_Y^{p+1}(X)) \longrightarrow \bigoplus_{\substack{m \geq 0 \\ p \geq 0}} K_m(\mathcal{M}_Y^p(X))$$

$$\nwarrow \qquad \swarrow$$

$$\bigoplus_{\substack{m \geq 0 \\ p \geq 0}} K_m(\mathcal{M}_Y^p(X)/\mathcal{M}_Y^{p+1}(X))$$

which gives us the desired spectral sequence $\{E_{rY}^{p,q}(X)\}$ converging to $K_{-p-q}(\mathcal{M}_Y(X)) = K_{-p-q}'^Y(X)$.   $\square$

From [Q1] we also know

$$K_m(\mathcal{M}_Y^p(X)/\mathcal{M}_Y^{p+1}(X)) \cong \bigoplus_{x \in X^{(p)} \cap Y} K_m(k(x)),$$

where $k(x)$ denotes the residue field of $x$; hence we have

$$E_{1Y}^{p,q}(X) \cong \begin{cases} \displaystyle\bigoplus_{x \in X^{(p)} \cap Y} K_{-p-q}(k(x)) & p \geq 0, p+q \leq 0 \\ 0 & \text{otherwise.} \end{cases}$$

We can also compute $E_{2Y}^{p,-p}(X)$. By Theorem 6 we get

$$d_1 = d_1^{p-1,-p} : E_{1Y}^{p-1,-p}(X) \longrightarrow E_{1Y}^{p,-p}(X) \longrightarrow 0.$$

We have

$$E_{1Y}^{p-1,-p}(X) \cong \bigoplus_{y \in X^{(p-1)} \cap Y} K_1(k(y)) \cong \bigoplus_{y \in X^{(p-1)} \cap Y} k(y)^*$$

and

$$E_{1Y}^{p,-p}(X) \cong \bigoplus_{x \in X^{(p)} \cap Y} K_0(k(x)) \cong \bigoplus_{x \in X^{(p)} \cap Y} \mathbb{Z} \cong Z_Y^p(X).$$

Hence we obtain

$$d_1 : \bigoplus_{y \in X^{(p-1)} \cap Y} k(y)^* \longrightarrow Z_Y^p(X) \longrightarrow 0.$$

One can show that for $(f_y) \in \bigoplus k(y)^*$ the image $d_1(f_y)$ is given by $\sum_y \operatorname{div} f_y \in Z_Y^p(X)$. We conclude that coker $d_1 = Z_Y^p(X)/\operatorname{im} d_1 = CH_Y^p(X)$, i.e.

$$E_{2Y}^{p,-p}(X) \cong CH_Y^p(X).$$

Similarly we compute $E_{2Y}^{p-1,-p}(X)$. Again by Theorem 6 we have

$$d_1 = d_1^{p-2,-p} : E_{1Y}^{p-2,-p}(X) \longrightarrow E_{1Y}^{p-1,-p}(X).$$

We know that $E_{1Y}^{p-2,-p}(X) \cong \bigoplus_{z \in X^{(p-2)} \cap Y} K_2(k(z))$. Hence we obtain

$$d_1 : \bigoplus_{z \in X^{(p-2)} \cap Y} K_2(k(z)) \longrightarrow \bigoplus_{y \in X^{(p-1)} \cap Y} k(y)^*.$$

The map $d_1 : K_2(k(z)) \longrightarrow k(y)^*$ is zero, unless $y \in \overline{\{z\}} =: Z$. To describe it in the latter case we first assume that the local ring $\mathcal{O}_{Z,y}$ is regular, hence a discrete valuation ring with valuation $v$, quotient field $k(z)$ and residue field $k(y)$. The map $d_1$ is then given by the *tame symbol* $\partial_v$ (see [G2]):

$$d_1(\{f,g\}) = \partial_v(\{f,g\})$$

$$:= \text{class of } ((-1)^{v(f)v(g)} \cdot f^{v(g)} \cdot g^{-v(f)}), \text{ if } f, g \in k(z)^* \backslash \{1\}.$$

In the non-regular case, let $W := \operatorname{Spec}(\widetilde{\mathcal{O}}_{Z,y})$ with $\widetilde{\mathcal{O}}_{Z,y}$ the normalization of $\mathcal{O}_{Z,y}$ (inside of $k(z)$) and denote by $y_1, \ldots, y_\ell$ the preimages of $y$ in $W$. Because $\mathcal{O}_{W,y_i}$ are now regular local rings, we can define $d_1$ by using the tame symbols $\partial_{v_i}$ (here $v_i$ denotes the discrete valuation of $\mathcal{O}_{W,y_i}$) and then the following formula holds :

$$d_1(\{f,g\}) := \prod_{i=1}^{\ell} N_{k(y_i)/k(y)} \partial_{v_i}(\{f,g\}),$$

where $k(y_i)$ denotes the residue field of $\mathcal{O}_{W,y_i}$ and $N_{k(y_i)/k(y)}$ is the norm from $k(y_i)$ to $k(y)$, a finite extension.

Let us define
$$CH_Y^{p-1,p}(X) := E_{2Y}^{p-1,-p}(X)$$

$$\cong \frac{\{(f_y) \in \bigoplus_{y \in X^{(p-1)} \cap Y} k(y)^* : \sum_y \operatorname{div} f_y = 0\}}{\{d_1(\{u_z\}) : (\{u_z\}) \in \bigoplus_{z \in X^{(p-2)} \cap Y} K_2(k(z))\}}.$$

**6.4**   In the previous section we observed that $E_{2Y}^{p,-p}(X) \cong CH_Y^p(X)$. We shall now sketch the proof of an isomorphism
$$E_{2Y}^{p,-p}(X)_{\mathbb{Q}} \cong Gr^p K_0^Y(X)_{\mathbb{Q}},$$

when $X$ is regular. This will finish the proof of Theorem 3(ii).

For this purpose, we first note that Quillen's $K$-groups $K_m'^Y(X)$ can be computed as follows [BG]. Denote by $\mathcal{K}$ the Zariski simplicial sheaf associated to the presheaf
$$U \mapsto \mathbb{Z} \times \varinjlim_n BGL_n(\Gamma(U, \mathcal{O}_X))^+ =: \mathbb{Z} \times BGL(\Gamma(U, \mathcal{O}_X))^+.$$

By means of flasque resolutions for Zariski simplicial sheaves, one can define a cohomology theory with coefficients in any such sheaf (loc. cit.). Applying this to $\mathcal{K}$ we define
$$K_m^Y(X) := H_Y^{-m}(X, \mathcal{K}).$$

When $X$ is regular, a generalization of Lemma 4 gives an isomorphism
$$K_m'^Y(X) \cong K_m^Y(X).$$

This definition of $K_m^Y(X)$ can be used to define operations
$$\lambda^k : K_m^Y(X) \to K_m^Y(X),$$

making $\bigoplus_{\substack{m \geq 0 \\ Y \subseteq X}} K_m^Y(X)$ into a $\lambda$-ring, where the product on this group is defined to be zero if both factors have positive degree.

More precisely, for all integers $k$, $0 \leq k \leq n$, exterior powers give maps of sheaves
$$\lambda^k : GL_n(\mathcal{O}_X) \longrightarrow GL_{\binom{n}{k}}(\mathcal{O}_X).$$

They can be used to construct maps of simplicial sheaves
$$\lambda^k : \mathbb{Z} \times BGL_n(\mathcal{O}_X)^+ \longrightarrow \mathcal{K},$$

which are stable for the inclusion $GL_n \hookrightarrow GL_{n+1}$, hence induce operations
$$\lambda^k : H_Y^{-m}(X, \mathcal{K}) \longrightarrow H_Y^{-m}(X, \mathcal{K}),$$

cf. [S1], Proposition 4. Furthermore, one can prove the following ([S1], Théorème 4):

**Theorem 7**   *When $X$ is regular the above constructions induce operations*

$$\lambda^k : E_{rY}^{p,q}(X) \longrightarrow E_{rY}^{p,q}(X)$$

*on the spectral sequence of Theorem 6. These converge to the operations*

$$\lambda^k : K_{-p-q}^Y(X) \longrightarrow K_{-p-q}^Y(X).$$

*For the isomorphism*

$$i_* : \bigoplus_{x \in X^{(p)} \cap Y} K_{-p-q}(k(x)) \xrightarrow{\cong} E_{1Y}^{p,q}(X)$$

*the Adams operations $\psi^k$, associated to these $\lambda^k$, satisfy*

$$\psi^k(i_*(\alpha)) = k^p \cdot i_*(\psi^k(\alpha))$$

*for any $\alpha \in \bigoplus_{x \in X^{(p)} \cap Y} K_{-p-q}(k(x))$.*

Because the operations $\psi^k$ act on $K_0$ (resp. $K_1$) of a field by the identity (resp. multiplication by $k$), the above Theorem implies that the operations $\psi^k$ act on $E_{rY}^{p,q}(X)$ by multiplication by $k^{-q}$ when $q = -p$ and $-p - 1$. Since the differentials

$$d_r^{p-1,-p} : E_{rY}^{p-1,-p}(X) \longrightarrow E_{rY}^{p-1+r,-p-r+1}(X)$$

commute with the Adams operations $\psi^k$, we have the relation $(k^{r-1} - 1) \cdot d_r^{p-1,-p} = 0$. Hence, if $r \geq 2$, after tensoring with $\mathbb{Q}$, $d_r^{p-1,-p}$ vanishes. This gives

$$E_{2Y}^{p,-p}(X)_{\mathbb{Q}} = E_{\infty Y}^{p,-p}(X)_{\mathbb{Q}} \cong Gr^p \, K_0^Y(X)_{\mathbb{Q}}.$$

So we finally get the desired isomorphism

$$CH_Y^p(X)_{\mathbb{Q}} \cong E_{2Y}^{p,-p}(X)_{\mathbb{Q}} \cong Gr^p \, K_0^Y(X)_{\mathbb{Q}}.$$

### 6.5

**Remarks**   For a bound on the denominators in the isomorphism above see [S1].

Gersten conjectured that, for any regular noetherian finite- dimensional scheme $X$ and for all $p \geq 0$, the group $CH^p(X)$ is isomorphic to the Zariski cohomology group $H^p(X, \mathcal{K}_p)$, where $\mathcal{K}_p$ is the sheaf of abelian groups attached to the presheaf $U \mapsto K_p(U)$; see [Q1]. In that case an intersection pairing

$$CH^p(X) \otimes CH^q(X) \longrightarrow CH^{p+q}(X)$$

could be defined as the cup product (up to sign)

$$H^p(X, \mathcal{K}_p) \otimes H^q(X, \mathcal{K}_q) \longrightarrow H^{p+q}(X, \mathcal{K}_{p+q})$$

induced by the products on higher $K$-groups; this definition is valid when $X$ is of finite type over some field, see for instance [G3].

# II

---

# Green Currents

In this chapter, following [GS2], we shall deal with some constructions in complex analytic geometry. If $X$ is a smooth projective complex variety, and $Z \subset X$ a closed irreducible subvariety, or more generally a cycle on $X$, we define a *Green current for $Z$* to be a current $g$ on $X$ such that $dd^c g + \delta_Z$ is a smooth form $\omega_Z$, where $\delta_Z$ denotes the current given by integration on $Z$. This notion generalizes the Green functions on a Riemann surface which were used by Arakelov to get an intersection theory on arithmetic surfaces [A1].

In the first section we prove the existence of these Green currents by applying the $\partial\bar{\partial}$-Lemma of Hodge theory (Theorem 1). When $Z$ has codimension one, a formula due to Poincaré and Lelong proves that these currents can be obtained using hermitian line bundles (Theorem 2). In general, we prove in the second paragraph that there exists a Green current for $Z$ given by a smooth form on $X - Z$ which is integrable on $X$ and grows slowly along $Z$. More precisely, it is of *logarithmic type along $Z$*; see 2.1 for the precise definition. The construction relies upon Hironaka's resolution of singularities to get to the case of divisors.

Given two cycles $Y$ and $Z$ on $X$ meeting properly, and $g_Y$ (resp. $g_Z$) a Green current for $Y$ (resp. $Z$), we then look for a Green current for the intersection cycle of $Y$ and $Z$ in $X$. The heuristic formula

$$(1) \qquad g_Y * g_Z := g_Y \wedge \delta_Z + \omega_Y \wedge g_Z$$

provides one since, formally, we have

$$dd^c(g_Y * g_Z) = dd^c(g_Y) \wedge \delta_Z + \omega_Y \wedge dd^c(g_Z)$$

$$= (\omega_Y - \delta_Y) \wedge \delta_Z + \omega_Y \wedge (\omega_Z - \delta_Z)$$

(2)

$$= -\delta_Y \wedge \delta_Z + \omega_Y \wedge \omega_Z$$

$$= -\delta_{Y \cap Z} + \omega_Y \wedge \omega_Z.$$

But (1) requires a rigorous definition of the product of currents $g_Y \wedge \delta_Z$ and the use of the rules of differentiation in (2) has to be justified. We show in the third section that this can be done when $g_Y$ is of logarithmic type (Theorem 4). We also quote from [GS2] several properties of this *-product.

## 1. Currents

**1.1**  Let $X$ be a smooth projective complex equidimensional variety of complex dimension $d$. Denote by $A^{p,q}(X)$ the vector space of $\mathbb{C}$-valued differential forms of type $(p,q)$. If $z = (z_1, \ldots, z_d)$ are local coordinates, we have, for $\omega \in A^{p,q}(X)$,

(3)

$$\omega = \sum_{\substack{1 \le i_1 < \ldots < i_p \le d \\ 1 \le j_1 < \ldots < j_q \le d}} f_{i_1 \cdots i_p, j_1 \cdots j_q}(z, \bar{z}) dz_{i_1} \wedge \ldots \wedge dz_{i_p} \wedge d\bar{z}_{j_1} \wedge \ldots \wedge d\bar{z}_{j_q}$$

$$= \sum_{\substack{|I|=p \\ |J|=q}} f_{I,J}(z, \bar{z}) dz_I \wedge d\bar{z}_J,$$

where $f_{I,J}(z, \bar{z})$ are $C^\infty$-functions. The space $A^n(X)$ of differential forms of degree $n$ is then given by

(4)
$$A^n(X) = \bigoplus_{p+q=n} A^{p,q}(X).$$

We denote by $\partial : A^{p,q}(X) \to A^{p+1,q}(X)$, $\bar{\partial} : A^{p,q}(X) \to A^{p,q+1}(X)$ and $d = \partial + \bar{\partial} : A^n(X) \to A^{n+1}(X)$ the usual differentials.

Following De Rham [DR1], we let $D_n(X) := A^n(X)^*$, i.e. $D_n(X)$ denote the space of linear functionals on $A^n(X)$, which are continuous in the sense of Schwartz: for a sequence $\{\omega_r\} \subset A^n(X)$ with $\mathrm{Supp}\,\omega_r$ contained in some fixed compact set $K$ and $T \in D_n(X)$, we have $T(\omega_r) \to 0$, if $\omega_r \to 0$, meaning that all coefficients in (3) together with finitely many of their derivatives tend uniformly to zero on $K$ when $r \to \infty$. By (4) we obtain the decomposition

$$D_n(X) = \bigoplus_{p+q=n} D_{p,q}(X),$$

$D_{p,q}(X)$ being the duals of $A^{p,q}(X)$.

**Definition 1** $D^{p,q}(X) := D_{d-p,d-q}(X)$.

**1.2** The differentials $\partial, \bar{\partial}, d$ induce maps from $D^{p,q}(X)$ to $D^{p+1,q}(X)$, $D^{p,q+1}(X)$, and $D^{p+q+1}(X)$ respectively. We denote these maps by $\partial', \bar{\partial}', d'$ respectively. For instance, if $T \in D^{p,q}(X)$, $\alpha \in A^{d-(p+1),d-q}(X)$ we have $(\partial' T)(\alpha) = T(\partial \alpha)$. Now we have the inclusion map

$$A^{p,q}(X) \longrightarrow D^{p,q}(X)$$
$$\omega \mapsto [\omega],$$

defined by

$$[\omega](\alpha) := \int_X \omega \wedge \alpha \qquad \text{for any } \alpha \in A^{d-p,d-q}(X).$$

Here we chose an orientation on $X$ by deciding that $dz_1 d\bar{z}_1 \cdots dz_n d\bar{z}_n$ has positive orientation on $\mathbb{C}^n$. If $p + q = n$, Stokes' Theorem leads to

$$
\begin{aligned}
[d\omega](\alpha) &= \int_X d\omega \wedge \alpha \\
&= \int_X d(\omega \wedge \alpha) - \int_X (-1)^n \omega \wedge d\alpha \\
&= (-1)^{n+1} \int_X \omega \wedge d\alpha \\
&= (-1)^{n+1} [\omega](d\alpha) \\
&= (-1)^{n+1} (d'[\omega])(\alpha).
\end{aligned}
$$

Let us denote $(-1)^{n+1}\partial'$, $(-1)^{n+1}\bar{\partial}'$, and $(-1)^{n+1}d'$ by $\partial, \bar{\partial}$ and $d$ respectively. We have commutative diagrams

$$
\begin{array}{ccc}
A^{p,q}(X) & \hookrightarrow & D^{p,q}(X) \\
\downarrow \partial & & \downarrow \partial \\
A^{p+1,q}(X) & \hookrightarrow & D^{p+1,q}(X)
\end{array}
$$

(and *idem* with $\bar{\partial}$ and $d$). These diagrams induce isomorphisms on the level of cohomology with respect to $\partial, \bar{\partial}$, and $d$, cf. [GH], p. 382.

Now, for every irreducible analytic subvariety $Y \overset{i}{\hookrightarrow} X$ of codimension $p$, we define a current $\delta_Y \in D^{p,p}(X)$ by setting, for all $\alpha \in A^{d-p,d-p}(X)$,

$$\delta_Y(\alpha) := \int_{Y^{ns}} i^* \alpha$$

where $Y^{ns}$ is the non-singular locus of $Y$. By linearity we extend this definition to arbitrary analytic subvarieties $Y$. The fact that $\delta_Y$ is well-defined and gives a current is due to Lelong [Le].

This also follows from Hironaka's theorem [Hi] on the resolution of singularities. This states that given any $Z \subset Y$, where $Z$ contains the singular locus of $Y$, there exists a proper map $\pi : \tilde{Y} \to Y$ such that

(i)  $\tilde{Y}$ is smooth;
(ii)  $E := \pi^{-1}(Z)$ is a divisor with normal crossings;
(iii)  $\pi : \tilde{Y} - E \to Y - Z$ is an isomorphism.

We then have

$$\delta_Y(\alpha) = \int_{Y^{ns}} i^*\alpha = \int_{\tilde{Y}} \pi^* i^*\alpha.$$

**1.3**   Let $d^c := (4\pi i)^{-1}(\partial - \bar{\partial})$  (hence $dd^c = -(2\pi i)^{-1}\partial\bar{\partial}$).

**Definition 2**   *A Green current for a codimension $p$ analytic subvariety $Y$, is a current $g \in D^{p-1,p-1}(X)$ such that*

$$dd^c g + \delta_Y = [\omega]$$

*for some form $\omega \in A^{p,p}(X)$.*

**Theorem 1**   *If $X$ is Kähler, then every $Y \subset X$ has a Green current. If $g_1, g_2$ are two Green currents for $Y$, then*

$$g_1 - g_2 = [\eta] + \partial S_1 + \bar{\partial} S_2$$

*with $\eta \in A^{p-1,p-1}(X)$, $S_1 \in D^{p-2,p-1}(X)$, $S_2 \in D^{p-1,p-2}(X)$.*

Before proving Theorem 1, we recall the Hodge decomposition theorem for $A^{p,q}(X)$ and $D^{p,q}(X)$. From a Kähler structure on $X$ we get adjoints $\partial^*, \bar{\partial}^*, d^*$ of $\partial, \bar{\partial}, d$ for the $L^2$-scalar product $\langle ., . \rangle$ on forms. The Laplacian is then defined as $\Delta_d := 2(\partial^*\partial + \partial\partial^*) = 2(\bar{\partial}^*\bar{\partial} + \bar{\partial}\bar{\partial}^*) = d^*d + dd^*$. If $\mathcal{H}^{p,q} := \ker\Delta_d$ denotes the harmonic forms, the Hodge decomposition for $A^{p,q}(X)$ are orthogonal decompositions

$$A^{p,q}(X) = \mathcal{H}^{p,q} \oplus \mathrm{im}\partial \oplus \mathrm{im}\partial^*$$
$$= \mathcal{H}^{p,q} \oplus \mathrm{im}\bar{\partial} \oplus \mathrm{im}\bar{\partial}^*$$
$$= \mathcal{H}^{p,q} \oplus \mathrm{im}d \oplus \mathrm{im}d^*.$$

By duality, completely analogous results hold for $D^{p,q}(X)$. As a consequence of this, we can prove the $\partial\bar{\partial}$-Lemma:

**Lemma 1** *If $T \in D^{p,p}(X)$ with $T = dS$ for some current $S$, then $T = dd^c U$ with $U \in D^{p-1,p-1}(X)$.*

*Proof.* We have $T = \partial S + \bar{\partial} S$; by Hodge decomposition we have

$$S = h_1 + \partial x_1 + \partial^* y_1 = h_2 + \bar{\partial} x_2 + \bar{\partial}^* y_2,$$
$$\partial S = \partial \bar{\partial} x_2 + \partial \bar{\partial}^* y_2,$$
$$\bar{\partial} S = \bar{\partial} \partial x_1 + \bar{\partial} \partial^* y_1,$$

hence

$$T = \bar{\partial} \partial x_1 + \partial \bar{\partial} x_2 + \partial \bar{\partial}^* y_2 + \bar{\partial} \partial^* y_1.$$

Now $\partial T = \bar{\partial} T = 0$, implies $\partial \bar{\partial} \partial^* y_1 = 0$, $\bar{\partial} \partial \bar{\partial}^* y_2 = 0$. Therefore, since $\partial^*$ anticommutes with $\bar{\partial}$ (see [GH]),

$$0 = \langle \partial \bar{\partial} \partial^* y_1, \bar{\partial} y_1 \rangle = -\langle \bar{\partial} \partial^* y_1, \bar{\partial} \partial^* y_1 \rangle.$$

So $\bar{\partial} \partial^* y_1 = 0$, and similarly $\partial \bar{\partial}^* y_2 = 0$. So we have

$$T = \partial \bar{\partial} (x_2 - x_1).$$

$\square$

*Proof of Theorem 1.* By Stokes' Theorem we have $d\delta_Y = 0$, hence by 1.2. we deduce $\delta_Y = [\omega] + dS$ for some $\omega \in A^{p,p}(X)$ and some current $S$. Now we deduce from Lemma 1 that

$$[\omega] - \delta_Y = -dS = dd^c g$$

for some $g \in D^{p-1,p-1}(X)$.

To prove the second part of the theorem, we first note $dd^c(g_1 - g_2) = [\omega_1] - [\omega_2] = dd^c[\eta']$ for some $\eta' \in A^{p-1,p-1}(X)$ by the $\partial\bar{\partial}$-Lemma for smooth currents, which is proved in the same way as Lemma 1. If we show that $dd^c x_0 = 0$, with $x_0$ of type $(p-1, p-1)$ implies $x_0 = [\eta''] + \partial S_1 + \bar{\partial} S_2$, the proof will be complete. But $\partial \bar{\partial} x_0 = 0$ implies $\bar{\partial} x_0 = [\eta_1] + \partial x_1$, hence $\bar{\partial} x_1 = [\eta_2] + \partial x_2$, because $\partial \bar{\partial} x_1 = \bar{\partial}[\eta_1]$. If we iterate this argument, we get a sequence of currents $x_n$ and forms $\eta_n$ such that $\bar{\partial} x_n = [\eta_{n+1}] + \partial x_{n+1}$, with $x_n$ of type $(p-n-1, q+n-1)$. When $n$ is big enough, we conclude that $\bar{\partial} x_n = [\eta_{n+1}]$, from which $x_n = [\xi_n] + \bar{\partial} v_n$ follows. Now $\bar{\partial} x_{n-1} = [\eta_n] + \partial x_n = [\eta_n] + \partial [\xi_n] + \partial \bar{\partial} v_n$ leads to $\bar{\partial}(x_{n-1} + \partial v_n) = [\eta_n] + \partial[\xi_n]$, hence $x_{n-1} = [\xi_{n-1}] + \partial u_{n-1} + \bar{\partial} v_{n-1}$, so, by repeating the argument, $x_0 = [\xi_0] + \partial u_0 + \bar{\partial} v_0$, as claimed. $\square$

## 1.4

**Theorem 2** (The Poincaré–Lelong formula). *Let $L$ be a holomorphic line bundle on $X$ with hermitian metric $\|\cdot\|$, $s$ a meromorphic section of*

$L$ and $c_1(L, \| \cdot \|)$ *the first Chern form of* $L$. *Then* $-\log \|s\|^2 \in L^1(X)$, *hence induces a distribution* $[-\log \|s\|^2] \in D^{0,0}(X)$. *This is a Green current for* div$s$:

$$dd^c[-\log \|s\|^2] + \delta_{\mathrm{div}s} = [c_1(L, \| \cdot \|)].$$

*Proof.*  By definition, an hermitian metric on $L$ is an hermitian scalar product on each fiber $L_x$, which varies smoothly with $x$; see also §IV.2.3. Since $L$ has rank one, this is the same as a smooth norm $\| \cdot \|$ on $L$. On $X$ minus the support of div$s$, one has an equality of differential forms

$$-dd^c \log \|s\|^2 = (2\pi i)^{-1} \partial\bar{\partial} \log \|s\|^2 = c_1(L, \| \cdot \|)$$

which can be taken as definition of the first Chern form $c_1(L, \| \cdot \|)$ on $X$, since replacing $s$ by $s' = f \cdot s$ with some non-vanishing holomorphic function $f$, gives

$$\partial\bar{\partial} \log \|s'\|^2 = \partial\bar{\partial} \log \|s\|^2,$$

because $\partial\bar{\partial} \log |f|^2 = 0$ (for another definition, see below Definition IV.2 ).

Using the theorem of resolution of singularities (see 1.2) we may assume that, in a local chart $U \simeq \mathbb{C}^n$, the divisor div$s$ has equation $z_1 \cdots z_k = 0$. By linearity we are reduced to the case $s = z_1$ and what we have to show is then the following equality (for every form $\omega$ of type $(n-1, n-1)$ with compact support in $U$):

$$\int_U \log |z_1|^2 dd^c \omega = \int_{z_1 = 0} \omega.$$

First we note

$$\int_U \log |z_1|^2 dd^c \omega = \lim_{\epsilon \to 0} \int_{|z_1| \geq \epsilon} \log |z_1|^2 dd^c \omega.$$

Now, applying Stokes' Theorem twice, we get

$$\lim_{\epsilon \to 0} \int_{|z_1| \geq \epsilon} \log |z_1|^2 \cdot dd^c \omega$$

$$= -\lim_{\epsilon \to 0} \int_{|z_1| = \epsilon} \log |z_1|^2 \cdot d^c \omega - \lim_{\epsilon \to 0} \int_{|z_1| \geq \epsilon} d\log |z_1|^2 \cdot d^c \omega$$

$$= \lim_{\epsilon \to 0} \int_{|z_1| \geq \epsilon} d^c \log |z_1|^2 \cdot d\omega$$

$$= \lim_{\epsilon \to 0} \int_{|z_1| = \epsilon} d^c \log |z_1|^2 \cdot \omega + \lim_{\epsilon \to 0} \int_{|z_1| \geq \epsilon} dd^c \log |z_1|^2 \cdot \omega$$

$$= \int_{z_1 = 0} \omega,$$

because $d^c \log |z_1|^2 = (4\pi i)^{-1}(\partial - \bar{\partial}) \log(z_1 \bar{z}_1) = (2\pi)^{-1} \mathrm{im}(dz_1/z_1) = (2\pi)^{-1} \cdot d(\theta_1)$ with $\theta_1 = \arg z_1$, and $dd^c \log |z_1|^2 = 0$. This completes the proof of the Poincaré–Lelong formula.  $\square$

## 2. Green forms of logarithmic type

**2.1**   As in the first section, let $X$ be a smooth projective complex variety and $Y$ an analytic subvariety of $X$ which does not contain any of its irreducible components.

**Definition 3**  *A smooth form $\alpha$ on $X - Y$ is said to be of logarithmic type along $Y$, if there exists a projective map $\pi : \widetilde{X} \to X$ such that $E := \pi^{-1}(Y)$ is a divisor with normal crossings, $\pi : \widetilde{X} - E \to X - Y$ is smooth and $\alpha$ is the direct image by $\pi$ of a form $\beta$ on $\widetilde{X} - E$ with the following property.*

*Near each $x \in \widetilde{X}$, let $z_1 z_2 ... z_k = 0$ be a local equation of $E$. Then there exists $\partial$ and $\bar{\partial}$ closed smooth forms $\alpha_i$ and a smooth form $\gamma$ such that*

$$(5) \qquad \beta = \sum_{i=1}^{k} \alpha_i \log |z_i|^2 + \gamma.$$

If $\alpha$ is of logarithmic type along $Y$, it is locally integrable on $X$, hence it defines a current $[\alpha]$, which is the direct image by $\pi$ of the current $[\beta]$.

**Lemma 2**

(i)   Let $f : X' \to X$ be a morphism of smooth projective varieties, and on $X - Y$, let $\alpha$ be a form of logarithmic type along the subvariety $Y$. If $f^{-1}(Y)$ does not contain any component of $X'$, the form $f^*(\alpha)$ is of logarithmic type along $f^{-1}(Y)$.

(ii)   Let $f : X \to X'$ be a projective morphism of smooth projective varieties, and let $\alpha$ be a form on $X - Y$ of logarithmic type along the subvariety $Y$. Assume that $f$ is smooth outside $Y$ and that $f(Y)$ does not contain any component of $X'$. Then $f_*(\alpha)$ is of logarithmic type along $f(Y)$ and $f_*([\alpha]) = [f_*(\alpha)]$.

*Proof.*  To prove (i) consider a diagram

$$
\begin{array}{ccc}
\widetilde{X'} & \xrightarrow{f'} & \widetilde{X} \\
\downarrow{\scriptstyle \pi'} & & \downarrow{\scriptstyle \pi} \\
X' & \xrightarrow{f} & X
\end{array}
$$

where $\pi : \tilde{X} \to X$ is as in Definition 3, $\pi'$ is projective, $(f\pi')^{-1}(Y)$ is a divisor with normal crossings in $\widetilde{X'}$, and the same diagram over $X - Y$ is cartesian. One gets $\widetilde{X'}$ by resolving the singularities of the Zariski closure of the fiber product of $X' - f^{-1}(Y)$ with $\tilde{X} - \pi^{-1}(Y)$ in the product of $X'$ and $\tilde{X}$ over $X$. If $\alpha = \pi_*(\beta)$, with $\beta$ as in (5), we get

$$f^*(\alpha) = f^*(\pi_*(\beta)) = \pi'_*(f'^*(\beta)).$$

Since $f'^*(\beta)$ satisfies the condition (5) on $\widetilde{X'}$, the assertion follows.

The proof of (ii) is similar; see [GS2] for more details.      □

## 2.2

**Theorem 3** *For every irreducible subvariety $Y \subset X$ there exists a smooth form $g_Y$ on $X - Y$ of logarithmic type along $Y$ such that $[g_Y]$ is a Green current for $Y$:*

$$dd^c[g_Y] + \delta_Y = [\omega]$$

*where $\omega$ is smooth on $X$.*

*Proof.*
*Step 1.*      Let us show the Theorem for a divisor $Y$ on $X$. In that case there exists a line bundle $L$ on $X$ with hermitian metric $\| \cdot \|$ and a section $s$ with $Y = \mathrm{div}\, s$. Putting $g_Y := -\log \|s\|^2$, we get by the Poincaré–Lelong formula

$$dd^c[g_Y] + \delta_Y = [c_1(L, \| \cdot \|)].$$

But $\log \|s\|^2$ is of logarithmic type along $Y$ as one can see by making $Y$ a divisor with normal crossings.

*Step 2.*      Let $f : Y \to X$ be a holomorphic map between complex manifolds of dimensions $d'$ and $d$ respectively. Then for the graph $\Gamma := \{(y, x) : x = f(y)\} \subset Y \times X$ there exists a Green form $g_\Gamma$ of logarithmic type along $\Gamma$.

To prove this, denote by $W$ the blow-up of $Y \times X$ along $\Gamma$. We have the following diagram:

$$
\begin{array}{ccc}
E & \xrightarrow{\ j\ } & W \\
\downarrow{\scriptstyle \pi_\Gamma} & & \downarrow{\scriptstyle \pi} \\
\Gamma & \xrightarrow{\ i\ } & Y \times X \\
{\scriptstyle \mathrm{pr}} \searrow & & \swarrow{\scriptstyle p} \quad {\scriptstyle q} \searrow \\
& Y & \qquad X
\end{array}
$$

with $E$ the exceptional divisor. The cohomology class $\mathrm{cl}(\Gamma)$ of $\Gamma$ is an element of $H_\Gamma^{d,d}(Y \times X, \mathbb{R})$, hence $\pi^* \mathrm{cl}(\Gamma) \in H_E^{d,d}(W, \mathbb{R}) \cong H^{d-1,d-1}(E, \mathbb{R})$.

**Lemma 3**    *There exists a closed form* $\alpha \in A^{d-1,d-1}(W)$ *such that* $\pi_*(\delta_E \wedge [\alpha]) = \delta_\Gamma$.

*Proof of Lemma 3.*    Here the currents $\delta_E \wedge [\alpha]$ and $\pi_*(\delta_E \wedge [\alpha])$ are defined by

$$\delta_E \wedge [\alpha](\omega) = \int_E j^* \alpha \wedge \omega$$

and

$$\pi_*(\delta_E \wedge [\alpha])(\omega) = \int_E j^* \alpha \wedge \pi^* \omega$$

for all $\omega$ of appropriate degree. By its very definition the exceptional divisor is the projective bundle

$$\begin{aligned}
E = \mathbb{P}(N_{Y \times X/\Gamma}) &\cong \mathbb{P}(i^* T_{Y \times X}/T_\Gamma) \\
&\cong \mathbb{P}(i^*(p^* T_Y \oplus q^* T_X)/T_\Gamma) \\
&\cong \mathbb{P}(i^* q^* T_X),
\end{aligned}$$

since $i^* p^* T_Y = p_\Gamma^* T_Y = T_\Gamma$; here $N_{Y \times X/\Gamma}$ denotes the normal bundle and $T_X, T_Y \ldots$ the tangent bundles of $X, Y \ldots$. Therefore $\bigoplus_{p \geq 0} H^{p,p}(E, \mathbb{R})$ is a free module over $\bigoplus_{p \geq 0} H^{p,p}(\Gamma, \mathbb{R})$ with basis $\xi^0, \xi^1, \ldots, \xi^{d-1}$ and with $\xi = j^* \mathrm{cl}(E)$ equal to the first Chern class of the tautological line bundle on $E$ ($\mathrm{cl}(E)$ being considered as an element of $H^{1,1}(W, \mathbb{R})$). Hence we have

$$\pi^* \mathrm{cl}(\Gamma) \cap \mathrm{cl}(W) = \sum_{i=0}^{d-1} \pi_\Gamma^*(a_i) \xi^i$$

with

$$a_i \in H^{d-1-i,d-1-i}(\Gamma, \mathbb{R}) \quad (i = 0, \ldots, d-1).$$

Now $\pi_\Gamma^* = j^* \pi^* p^* (p_\Gamma^*)^{-1}$, because $p_\Gamma \pi_\Gamma = p \pi j$ and

$$p_\Gamma^* : H^*(Y, \mathbb{R}) \to H^*(\Gamma, \mathbb{R})$$

is an isomorphism. Hence, putting

$$b_i := \pi^* \, p^* \, (p_\Gamma^*)^{-1}(a_i) \in H^{d-1-i,d-1-i}(W, \mathbb{R}),$$

we find

$$\pi^* \mathrm{cl}(\Gamma) \cap \mathrm{cl}(W) = \sum_{i=0}^{d-1} j^*(b_i) j^* \, \mathrm{cl}(E)^i = j^* \left( \sum_{i=0}^{d-1} b_i \cdot \mathrm{cl}(E)^i \right),$$

and if $\alpha \in A^{d-1,d-1}(W)$ represents $\sum_{i=0}^{d-1} b_i \cdot \mathrm{cl}(E)^i$ we have

$$\pi^* \mathrm{cl}(\Gamma) \cap \mathrm{cl}(W) = j^* \, \mathrm{cl}(\alpha).$$

Because $E$ is smooth over $\Gamma$, we can apply $\pi_{\Gamma_*}$ to the last identity (integration along the fibres) and we obtain by the projection formula

$$\pi_{\Gamma_*} j^* \; \mathrm{cl}(\alpha) = \pi_{\Gamma_*}(\pi^* \; \mathrm{cl}(\Gamma) \cap \mathrm{cl}(W)) =$$
$$\mathrm{cl}(\Gamma) \cap \pi_* \; \mathrm{cl}(W) = \mathrm{cl}(\Gamma) \cap \mathrm{cl}(Y \times X) = \mathrm{cl}(\Gamma) \in H^{0,0}(\Gamma, \mathbb{R}).$$

Now $\mathrm{cl}(\Gamma)$ induces the current $\delta_\Gamma$ and $\pi_{\Gamma_*} j^* \; \mathrm{cl}(\alpha)$ induces the current

$$[\pi_{\Gamma_*} j^* \; \alpha](\omega) = \pi_*[j^*\alpha](\omega) = \int_E j^*\alpha \wedge \pi^*\omega = \pi_*(\delta_E \wedge [\alpha])(\omega).$$

Because we are in degree 0, we conclude that

$$\pi_*(\delta_E \wedge [\alpha]) = \delta_\Gamma$$

and the lemma is proved.                                                    □

We will now construct the Green form $g_\Gamma$. Let $L$ be the line bundle on $W$ with hermitian metric $\| \cdot \|$ and a section $s$ with $E = \mathrm{div} s$. By Step 1, we know that $dd^c([\log \|s\|^2]) = -[\beta] + \delta_E$ (here $\beta = c_1(L, \| \cdot \|)$). Multiplying the last equation by the $\alpha$ constructed in Lemma 2, we get

$$dd^c[(\log \|s\|^2)\alpha] = -[\beta \wedge \alpha] + \delta_E \wedge [\alpha].$$

By Lemma 2, the current $\delta_E \wedge [\alpha]$ represents the cohomology class $\pi^*\mathrm{cl}(\Gamma) \in \dot{H}^{d,d}(W, \mathbb{R})$, hence, by the above equation, the same is true for $[\beta \wedge \alpha]$. Taking now $\omega \in A^{d,d}(Y \times X)$ such that $\pi^*\omega$ represents $\pi^*\mathrm{cl}(\Gamma)$, we know by the $\partial\bar{\partial}$-Lemma that there exists $\varphi \in A^{d-1,d-1}(W)$ satisfying

$$dd^c\varphi = \beta \wedge \alpha - \pi^*\omega.$$

With these ingredients we put

$$\widetilde{g}_\Gamma := -((\log \|s\|^2)\alpha + \varphi)$$

and denote by $g_\Gamma$ the form corresponding to $\widetilde{g}_\Gamma$ via the isomorphism $W - E \cong (Y \times X) - \Gamma$. Because

$$dd^c[(\log \|s\|^2)\alpha + \varphi] = -[\beta \wedge \alpha] + \delta_E \wedge [\alpha] + [\beta \wedge \alpha] - [\pi^*\omega],$$

we get by Lemma 2

$$dd^c[g_\Gamma] = -\pi_*(\delta_E \wedge [\alpha] - [\pi^*\omega]) = [\omega] - \delta_\Gamma.$$

Using Lemma 2(ii) we see that $g_\Gamma$ is of logarithmic type along $\Gamma$, which concludes the proof of Step 2.

*Step 3.* Let $X$ be a projective complex manifold and $i : Y \hookrightarrow X$ a closed submanifold of codimension $n$. Then for any closed form $\alpha \in A^{p,p}(Y)$ there exists a smooth form $g$ on $X - Y$ of type $(n + p - 1, n + p - 1)$ which is of logarithmic type along $Y$ and such that

$$dd^c[g] = [\beta] - i_*[\alpha]$$

for some $\beta \in A^{n+p,n+p}(X)$.

*Proof.*    Denote by $\Gamma \subset Y \times X$ the graph of $i : Y \hookrightarrow X$ and let $p : Y \times X \to Y$, $q : Y \times X \to X$ be the projections. Now consider a Green form $g_\Gamma$ constructed in Step 2. The form $g := q_*(g_\Gamma \wedge p^*\alpha)$ is smooth on $X - Y$ of type $(n+p-1, n+p-1)$, and of logarithmic type along $Y$, by Lemma 2(ii). We compute

$$
\begin{aligned}
dd^c[g] &= dd^c[q_*(g_\Gamma \wedge p^*\alpha)] \\
&= [q_* dd^c(g_\Gamma \wedge p^*\alpha)] \\
&= [q_*(dd^c g_\Gamma \wedge p^*\alpha)] \\
&= q_*(dd^c[g_\Gamma] \wedge [p^*\alpha]) \\
&= q_*(([\omega] - \delta_\Gamma) \wedge [p^*\alpha]) \\
&= [\beta] - i_*[\alpha].
\end{aligned}
$$

Here we took $\beta := q_*(\omega \wedge p^*\alpha) \in A^{n+p,n+p}(X)$ and observed that, for all $\eta$ of appropriate degree,

$$
\begin{aligned}
q_*(\delta_\Gamma \wedge [p^*\alpha])(\eta) &= (\delta_\Gamma \wedge [p^*\alpha])(q^*\eta) = \int_\Gamma p^*\alpha \wedge q^*\eta \\
&= \int_\Gamma p_\Gamma^*\alpha \wedge q^*\eta = \int_Y \alpha \wedge p_\Gamma^{*-1} q^*\eta \\
&= \int_Y \alpha \wedge i^*\eta = [\alpha](i^*\eta) \\
&= i_*[\alpha](\eta),
\end{aligned}
$$

with $p_\Gamma = p|\Gamma : \Gamma \xrightarrow{\cong} Y$, as in Step 2.

*Step 4.*    We can now prove Theorem 3. Let $i : Y \hookrightarrow X$ be an irreducible subvariety of $X$. Then, see §1.2, there exists a smooth projective complex variety $\widetilde{X}$ and a proper map $\pi : \widetilde{X} \to X$ such that $\widetilde{X} - E \cong X - Y$, where $E = \pi^{-1}(Y)$ is a divisor with normal crossings, i.e. $E = \bigcup_{i=1}^k E_i$ with $E_i$ smooth divisors such that for every $x \in E_{i_1} \cap \ldots \cap E_{i_\ell}$ there exist local coordinates $z_1, \ldots, z_m$ at $x$ such that $E_{i_j}$ is given by the equation $z_{i_j} = 0$, $1 \le j \le \ell$. Now consider the cycle class $[Y] \in CH_Y^n(X)$. We have $\pi^*[Y] \in CH_E^n(\widetilde{X}) \cong CH^{n-1}(E) = \bigoplus_i CH^{n-1}(E_i)$ (cf. [Fu]), hence $\pi^*[Y] = \sum_i[\eta_i]$ with $[\eta_i] \in CH^{n-1}(E_i)$. Therefore the cohomology class $\pi^* \mathrm{cl}(Y) \in H_E^{n,n}(\widetilde{X}, \mathbb{R})$ decomposes to $\sum_i j_{i*}(\mathrm{cl}(\alpha_i))$, where $\alpha_i$ is a closed form of $A^{n-1,n-1}(E_i)$ and $j_i : E_i \hookrightarrow E$ is the inclusion $(i = 1, \ldots, k)$. Now, by Step 3, there exist smooth forms $g_i$ on $\widetilde{X} - E_i$ of type $(n-1, n-1)$ with prescribed order of growth near $E_i$ satisfying

$$
dd^c[g_i] = [\beta_i] - j_{i*}[\alpha_i]
$$

for some $\beta_i \in A^{n,n}(\widetilde{X})$. Because $\pi^* \mathrm{cl}(Y) = \sum_i j_{i*}(\mathrm{cl}(\alpha_i)) = \sum_i \mathrm{cl}(\beta_i)$, there is a closed form $\omega \in A^{n,n}(X)$ such that $\pi^*\omega - \sum_i \beta_i$ is an exact form, hence, by the $\partial\bar{\partial}$-Lemma, there exists $\widetilde{g} \in A^{n-1,n-1}(\widetilde{X})$ with

$$dd^c\widetilde{g} = \pi^*\omega - \sum_i \beta_i.$$

We finally put

$$\widetilde{g}_Y := \sum_i g_i + \widetilde{g}$$

and denote by $g_Y$ the form corresponding to $\widetilde{g}_Y$ via the isomorphism $\pi : \widetilde{X} - E \cong X - Y$. The form $g_Y$ is a smooth form on $X - Y$ of logarithmic type along $Y$, by Step 3. Furthermore, it satisfies

$$dd^c[g_Y] = dd^c\Big(\sum_i \pi_*[g_i] + \pi_*[\widetilde{g}]\Big)$$

$$= \sum_i \pi_*([\beta_i] - j_{i*}[\alpha_i]) + \pi_*[\pi^*\omega] - \sum_i \pi_*[\beta_i]$$

$$= [\omega] - \sum_i \pi_* j_{i*}[\alpha_i].$$

To complete the proof of Theorem 3, we have to show

$$\delta_Y = \sum_i \pi_* j_{i*}[\alpha_i].$$

For this we look at the resolution of singularities $\widetilde{Z}_i$ of $Z_i := \pi(E_i) \subset X$. Consider the following diagram:

$$
\begin{array}{ccccc}
\widetilde{E}_i & \xrightarrow{\widetilde{\pi}_i} & \widetilde{Z}_i & & \\
q_i \downarrow & & \downarrow p_i & & \\
E_i & \xrightarrow{\pi_i} & Z_i & \xhookrightarrow{j} & X
\end{array}
$$

where $q_i$ is birational and $\widetilde{E}_i$ is smooth. We have the equation

$$\pi_* j_{i*}[\alpha_i] = j_* p_{i*} \widetilde{\pi}_{i*}[q_i^* \alpha_i].$$

If $\mathrm{codim}_X Z_i > \mathrm{codim}_X Y = n$, then $p_{i*}\widetilde{\pi}_{i*}[q_i^*\alpha_i] = 0$. If $\mathrm{codim}_X Z_i = \mathrm{codim}_X Y$, then $Z_i = Y$ because $Y$ is irreducible. Therefore

$$\sum_i \pi_* j_{i*}[\alpha_i] = p_*(S)$$

for $p : \widetilde{Y} \to Y$ a resolution of singularities of $Y$ and $S$ a closed current in $D^{0,0}(\widetilde{Y})$. So $S = \lambda \cdot \delta_{\widetilde{Y}}$, hence $\sum_i \pi_* j_{i*}[\alpha_i] = \lambda \cdot \delta_Y$ for some real number $\lambda$. But $\lambda = 1$, because $\delta_Y$ has cohomology class $\mathrm{cl}(Y) \in H^{p,p}(X,\mathbb{R})$, so the class of $\lambda\delta_Y$ is $\pi_*\pi^*\mathrm{cl}(Y) \in H^{p,p}(X,\mathbb{R})$, but $\pi_*\pi^*\mathrm{cl}(Y) = \mathrm{cl}(Y)$.

$\square$

**2.3**   Here are two examples of Green currents:

(1)   Let $X = \mathbb{C}^2$ (with coordinates $z_1, z_2$), and $Y = \{0\}$. Then the form

$$g_0 := \log(|z_1|^2 + |z_2|^2) \cdot dd^c \log(|z_1|^2 + |z_2|^2)$$

on $X - Y$ is locally integrable on $X$ and the associated current $[g_0]$ satisfies $dd^c[g_0] = -\delta_0$, hence $[g_0]$ defines a Green current on $X$ for $Y$. Its image by $\partial$ was introduced by Bochner and Martinelli [GH].

(2)   $X = \mathbb{P}^d$ (with homogeneous coordinates $X_0, \ldots, X_d$), $Y$ defined by $X_0 = \ldots = X_{p-1} = 0$. Define

$$\tau := \log(|X_0|^2 + \ldots + |X_d|^2), \quad \alpha := dd^c\tau \quad \text{on } X;$$

$$\sigma := \log(|X_0|^2 + \ldots + |X_{p-1}|^2), \quad \beta := dd^c\sigma \quad \text{on } X - Y;$$

$$\Lambda := (\tau - \sigma)\left(\sum_{i=0}^{p-1} \alpha^i \wedge \beta^{p-1-i}\right) \quad \text{on } X - Y.$$

By a theorem of H. Levine [Lv], the smooth form $\Lambda$ on $X - Y$ is integrable and the associated current on $X$ satisfies

$$dd^c[\Lambda] = [\alpha^p] - \delta_Y.$$

By blowing up $Y$ one can check that $\Lambda$ is of logarithmic type along $Y$ [GS3].

**2.4**   There are relatively few examples of explicit Green currents . For example, one would like to extend the formula of Levine given above to the case of arbitrary Schubert cells in the Grassmannian varieties. This problem was adressed by Stoll [St], but he did not get a canonical and explicit solution. We shall see below, in §IV.4, that a canonical Green current exists but we do not know any explicit formula for it.

**2.5**   The concept of Green currents plays a central role in Nevanlinna theory (see [Sh]). We finish this section with a very brief description of their use. Let $X$ be a smooth projective complex variety and $Y$ a closed analytic subset of codimension $p$. Let $f : \mathbb{C}^p \to X$ be a holomorphic map with $f(0) \notin Y$. The central question of Nevanlinna theory is the following: does $f(\mathbb{C}^p)$ meet $Y$?

Let $g_Y$ be a Green current for $Y$, i.e., $dd^c g = [\omega] - \delta_Y$. For $r \in \mathbb{R}^+$, define

$$N_f(r) := \int_0^r \frac{dt}{t} \#\{z \in \mathbb{C}^p : |z| < t, f(z) \in Y\}.$$

At least formally, we get

$$N_f(r) = \int_0^r \frac{dt}{t} \int_{|z| \le t} f^*(\delta_Y) = \int_0^r \frac{dt}{t} \int_{|z| \le t} f^*(\omega - dd^c g_Y),$$

hence by Stokes' Theorem

$$N_f(r) = T_f(r) + R_f(r) - \frac{1}{2} \int_{|z|=r} f^*(g_Y) d^c \log |z|^2 + O(1),$$

where

$$T_f(r) = \int_0^r \frac{dt}{t} \int_{|z| \le t} f^*(\omega), \quad R_f(r) = \frac{1}{2} \int_{|z| \le r} f^*(g_Y) dd^c \log |z|^2.$$

If $g_Y$ is positive, we obtain the estimate

$$N_f(r) \le T_f(r) + R_f(r) + O(1),$$

hence the behavior of $T_f(r)$ and $R_f(r)$ for $r \to \infty$ gives an asymptotic description of $N_f(r)$, hence of the number of times the image by $f$ of the ball of radius $r$ meets $Y$.

On the other hand, as we shall see in the next chapter, Green currents can be used to develop some intersection theory on arithmetic varieties. This fact might well give a key to explaining Vojta's analogy between Nevanlinna theory and arithmetic geometry, which led him to formulate far-reaching conjectures on diophantine equations [V1].

## 3. The ∗-product of Green currents

**3.1**    Let $X$ be as in Section 2 and let $Y, Z \subset X$ be closed irreducible subsets such that $Z \not\subset Y$. Denote by $g_Y$ a Green form of logarithmic type for $Y$, as constructed in Theorem 3. Let $p : \tilde{Z} \longrightarrow Z$ be a resolution of the singularities of $Z$ and $q : \tilde{Z} \longrightarrow X$ its composite with the inclusion of $Z$ into $X$. By Lemma 2(i) we know that $q^* g_Y$ is of logarithmic type along the inverse image of $Y$ in $\tilde{Z}$. In particular it is integrable and the formula

$$[g_Y] \wedge \delta_Z := q_*[q^* g_Y]$$

defines a current on $X$. For any Green current $g_Z$ for $Z$, we define the ∗-product with $[g_Y]$ to be

**Definition 4**    $[g_Y] * g_Z := [g_Y] \wedge \delta_Z + [\omega_Y] \wedge g_Z$. If $\mathrm{codim}_X Y = n$ and $\mathrm{codim}_X Z = m$, then $[g_Y] * g_Z$ lies in $D^{n+m-1,n+m-1}(X)$.

**Theorem 4**    *If* $Y, Z$ *intersect properly, i.e., if* $Y \cap Z = \bigcup_i S_i$ *with* $\mathrm{codim}_X S_i = \mathrm{codim}_X Y + \mathrm{codim}_X Z = n + m$, *then*

$$dd^c([g_Y] * g_Z) = [\omega_Y \wedge \omega_Z] - \sum_i \mu_i \delta_{S_i},$$

where the integers $\mu_i = \mu_i(Y, Z)$ are Serre's intersection multiplicities, defined in §I.2.

**Proof.** Since $q^* g_Y$ is of logarithmic type along $q^{-1}(Y)$, there exists a projective map $\pi : Z' \to \tilde{Z}$ such that $E := (q\pi)^{-1}(Y)$ is a divisor with normal crossings, $\pi : Z' - E \to \tilde{Z} - q^{-1}(Y)$ is smooth and $q^* g_Y$ is the direct image by $\pi$ of a form $\beta$ on $Z' - E$ which can be written locally

$$\beta = \sum_{i=0}^{k} \alpha_i \log |z_i|^2 + \gamma$$

as in (5). Using the fact that $\alpha_i$ is $\partial$ and $\bar{\partial}$ closed and the Poincaré-Lelong formula, we get

$$dd^c[\beta] = \sum_{i=0}^{k} j_{i*}[a_i] + [c],$$

where $a_i$ is smooth and closed on the $i$-th component $E_i$ of $E$, $j_i : E_i \to Z'$ denotes the inclusion, and $c$ is smooth on the whole of $Z'$. If $r$ is the composite of $\pi$ with $q$, we deduce

$$dd^c([g_Y] \wedge \delta_Z) = dd^c(q_*[q^* g_Y]) = q_* dd^c[q^* g_Y]$$

$$= \sum_{i=0}^{k} r_* j_{i*}[a_i] + r_*[c].$$

Let $\omega_Y$ be the smooth form on $X$ such that $dd^c[g_Y] + \delta_Y = [\omega_Y]$. By restricting this equation outside $Y$ we get

$$r_*[c] = q_*[q^*(\omega_Y)] = [\omega_Y] \wedge \delta_Z.$$

On the other hand, since the map $r : E_i \to r(E_i)$ is generically smooth ([H], Corollary 10.7) the Fubini Study theorem implies that there is an integrable form $b_i$ on $r(E_i)$ whose associated current on $X$ is closed and equal to $r_* j_{i*}[a_i]$. If the codimension of $r(E_i)$ is bigger than $m + n$, this current must vanish since the degree of $a_i$ is too small. When the codimension of $r(E_i)$ is equal to $m + n$, the form $b_i$ is closed of degree zero, hence it is a constant $\lambda_i$. So we conclude that

(6)                    $$dd^c([g_Y] * g_Z) = [\omega_Y \wedge \omega_Z] - R,$$

with $R = \sum_i \lambda_i \delta_{S_i}$.

To show the equality $\lambda_i = \mu_i(Y, Z)$, we use the cohomology with supports. First we have

$$\mathrm{cl}(Y) \in H_Y^{2n}(X) \cong H^0(Y),$$
$$\mathrm{cl}(Z) \in H_Z^{2m}(X) \cong H^0(Z),$$

hence

$$\mathrm{cl}(Y) \cdot \mathrm{cl}(Z) \in H^{2n+2m}_{Y \cap Z}(X) \cong H^0(Y \cap Z) \cong \bigoplus_i H^0(S_i)$$

$$\cong \bigoplus_i H^{2n+2m}_{S_i}(X)$$

and

$$\mathrm{cl}(Y) \cdot \mathrm{cl}(Z) = \sum_i \mu_i(Y, Z) \cdot \mathrm{cl}(S_i).$$

If we show that $\mathrm{cl}(R)$ is equal to $\mathrm{cl}(Y) \cdot \mathrm{cl}(Z)$, we obtain, since $\mathrm{cl}(\delta_{S_i})$ equals $\mathrm{cl}(S_i)$, that $\lambda_i = \mu_i(Y, Z)$. To prove the last claim, we view the cohomology of $X$ with support in $Y$ as the cohomology of the cone of the restriction morphism $D^*(X) \longrightarrow D^*(X - Y)$. In this complex $\mathrm{cl}(Y)$ can be represented by the cycle

$$(\delta_Y, 0) \in D^{2n}(X) \oplus D^{2n-1}(X - Y),$$

which is cohomologous to $([\omega_Y], d^c[g_Y])$. Similarly $\mathrm{cl}(Z)$ is represented by

$$(\delta_Z, 0) \in D^{2m}(X) \oplus D^{2m-1}(X - Z).$$

Therefore, $\mathrm{cl}(Y) \cdot \mathrm{cl}(Z)$ is represented by

$$([\omega_Y] \wedge \delta_Z, d^c[g_Y] \wedge \delta_Z) = (\delta_Z \wedge [\omega_Y], d^c[g_Y] \wedge \delta_Z),$$

which is cohomologous to

$$(R, 0) \in D^{2n+2m}(X) \oplus D^{2n+2m-1}(X - (Y \cap Z))$$

by equation (6); see [GS2] §2.1 for more details. The theorem follows.
□

**3.2**    Let $Y \subset X$ be as above, let $Z$ be an irreducible smooth projective complex variety and $f : Z \to X$ a map with $f^{-1}(Y) \neq Z$. If $g_Y$ is a Green form of logarithmic type for $Y$ its pull-back $f^*(g_Y)$ is of logarithmic type along $f^{-1}(Y)$ (Lemma 2 (i)), so we define a current

$$f^*[g_Y] := [f^* g_Y] \in D^{n-1,n-1}(Z).$$

If $f^{-1}(Y) = \bigcup_i S_i$ with $\mathrm{codim}_Z S_i = \mathrm{codim}_X Y$, then one can show, by the same proof as Theorem 4, that

$$dd^c f^*[g_Y] = [f^* \omega_Y] - \sum_i \mu_i \delta_{S_i},$$

where $\mu_i$ are the multiplicities of the cycle $f^*(Y)$.

**3.3**    Let $Y \subset X$ be a closed irreducible subset and $g_Y$ a Green current for $Y$. By Theorem 3, there exists a Green form $\tilde{g}_Y$ of logarithmic type for $Y$. By Theorem 1, we have

$$g_Y = [\tilde{g}_Y] + [\eta] + \partial S_1 + \bar{\partial} S_2,$$

whence every Green current for $Y$ may be represented by a Green form of logarithmic type along $Y$ modulo $(\mathrm{im}\,\partial + \mathrm{im}\,\bar{\partial})$.

Let $Y, Z \subset X$ be closed irreducible subsets such that $Z \not\subset Y$, and $g_Y$ (resp. $g_Z$) a Green current for $Y$ (resp. $Z$). We define the ∗-product of $g_Y$ with $g_Z$ by

$$g_Y * g_Z := [\tilde{g}_Y] * g_Z \quad \text{modulo } (\mathrm{im}\,\partial + \mathrm{im}\,\bar{\partial})$$

where $\tilde{g}_Y$ is any Green form of logarithmic type for $Y$ congruent to $g_Y$ modulo $\mathrm{im}\,\partial + \mathrm{im}\,\bar{\partial}$, and $[\tilde{g}_Y] * g_Z$ is defined as in 3.1. One can show that this definition does not depend on the choice of $\tilde{g}_Y$.

Furthermore, the ∗-product is commutative, i.e.

$$g_Y * g_Z = g_Z * g_Y \quad \text{modulo } (\mathrm{im}\,\partial + \mathrm{im}\,\bar{\partial}).$$

It is also associative. If $Y, Z, W \subset X$ are closed irreducible subsets meeting properly and $g_Y$ (resp. $g_Z, g_W$) Green currents for $Y$ (resp. $Z, W$), then we have

$$g_Y * (g_Z * g_W) = (g_Y * g_Z) * g_W \quad \text{modulo } (\mathrm{im}\,\partial + \mathrm{im}\,\bar{\partial}).$$

The proof of these facts in full generality is technical. It uses the precise form of Hironaka's theorem about resolution of singularities ([Hi], Theorem II), see [GS2]. Let us just notice here that, if we compute formally with currents as if they were forms, we get

$$
\begin{aligned}
g_Y * g_Z &= g_Y \wedge \delta_Z + \omega_Y \wedge g_Z \\
&= g_Y \wedge \delta_Z + \delta_Y \wedge g_Z + dd^c g_Y \wedge g_Z \\
&= g_Y \wedge \delta_Z + \delta_Y \wedge g_Z + g_Y \wedge dd^c g_Z \\
&= g_Z * g_Y,
\end{aligned}
$$

and

$$
\begin{aligned}
g_Y * (g_Z * g_W) &= g_Y \wedge \delta_Z \wedge \delta_W + \omega_Y \wedge g_Z \wedge \delta_W + \omega_Y \wedge \omega_Z \wedge g_W \\
&= (g_Y * g_Z) * g_W.
\end{aligned}
$$

# III

---

# Arithmetic Chow Groups

In this chapter, following [GS2] (see also [G1] for an introduction), we begin to develop Arakelov geometry in higher dimensions. Given a regular projective flat scheme $X$ over $\mathbb{Z}$, we introduce in §1 its arithmetic Chow groups $\widehat{CH}^p(X)$, which are generated by pairs $(Z, g_Z)$, where $Z$ is a cycle of codimension $p$ on $X$ , and $g_Z$ is a Green current for the corresponding cycle on the set $X(\mathbb{C})$ of complex points of $X$. We prove in Theorem 1 that these groups sit in several exact sequences. In §2, we use our previous study of intersection theory on regular schemes Chapter I, and $*$-products of Green currents, in Chapter II, to get an intersection pairing

$$\widehat{CH}^p(X) \otimes \widehat{CH}^q(X) \longrightarrow \widehat{CH}^{p+q}(X)_{\mathbb{Q}}.$$

In §3 we show that $\widehat{CH}^p(X)$ is contravariant in $X$, and covariant for maps which are smooth on the generic fiber. After giving a few examples in §4, we assume in §5 that $X(\mathbb{C})$ is endowed with a Kähler metric. It is then possible to define a subgroup of $\widehat{CH}^p(X)$ by imposing harmonicity conditions; when $p = 1$ and $X$ is an arithmetic surface, we recover Arakelov's definitions [A1], [A2]. Finally, in §6, we describe, without proofs, recent results on the height of projective varieties where the formalism of arithmetic intersection has been used.

## 1. Definitions

**1.1**   In this chapter, $X$ will always be a regular scheme, projective and flat over Spec $\mathbf{Z}$. Such a scheme will be called an *arithmetic variety*. Denote complex conjugation by $F_\infty : X(\mathbb{C}) \to X(\mathbb{C})$; it is a continuous involution of $X(\mathbb{C})$. We put

$$A^{p,p}(X) := \{\omega \in A^{p,p}(X(\mathbb{C})) : \omega \text{ real}, F_\infty^* \omega = (-1)^p \omega\}$$

$$Z^{p,p}(X) := \ker(d : A^{p,p}(X) \longrightarrow A^{2p+1}(X(\mathbb{C}))) \subset A^{p,p}(X)$$

$$H^{p,p}(X) := \{c \in H^{p,p}(X(\mathbb{C})) : c \text{ real}, F_\infty^* c = (-1)^p c\}$$

$$\widetilde{A}^{p,p}(X) := A^{p,p}(X)/(\mathrm{im}\,\partial + \mathrm{im}\,\overline{\partial})$$

$$D^{p,p}(X) := \{T \in D^{p,p}(X(\mathbb{C})) : T \text{ real}, F_\infty^* T = (-1)^p T\}.$$

We note that $H^{p,p}(X) = \ker dd^c/(\mathrm{im}\,\partial + \mathrm{im}\,\overline{\partial}) \subset \widetilde{A}^{p,p}(X)$ because $X(\mathbb{C})$ is projective and hence Kähler. As in §II.1.2, we have an imbedding $A^{p,p}(X) \hookrightarrow D^{p,p}(X)$ mapping $\omega$ to $[\omega]$. Furthermore, any cycle $Z = \sum_\alpha Z_\alpha \in Z^p(X)$ defines a current $\delta_Z = \sum_\alpha \delta_{Z_\alpha} \in D^{p,p}(X)$.

A *Green current* for such a cycle is an element $g_Z \in D^{p-1,p-1}(X)$ such that there exists a form $\omega_Z \in A^{p,p}(X)$ for which

$$dd^c g_Z + \delta_Z = [\omega_Z]$$

in $D^{p,p}(X)$.

For example, let $y$ be a point in $X^{(p-1)}$, i.e. $Y = \overline{\{y\}}$ is a closed integral subscheme of $X$ of codimension $p-1$. Then any rational function $f \in k(y)^*$ on $Y$ induces a rational function $f_\infty$ on $Y(\mathbb{C})$. If $\widetilde{Y}$ is a resolution of singularities of $Y(\mathbb{C})$, $f_\infty$ restricts to a rational function $\widetilde{f}_\infty$ on $\widetilde{Y}$. Hence $\log |\widetilde{f}_\infty|^2$ is a real-valued $L^1$-function on $\widetilde{Y}$ and therefore defines a current in $D^{0,0}(\widetilde{Y})$, whence $\widetilde{i}_*[\log |\widetilde{f}_\infty|^2] \in D^{p-1,p-1}(X)$, where $\widetilde{i}$ is the natural map $\widetilde{Y} \to X(\mathbb{C})$. We denote this current by $[\log |f|^2]$, i.e.

$$[\log |f|^2](\omega) = \widetilde{i}_*[\log |\widetilde{f}_\infty|^2](\omega) = \int_{\widetilde{Y}} \log |\widetilde{f}_\infty|^2 \cdot \widetilde{i}^*(\omega) .$$

By the Poincaré–Lelong formula (see Theorem II.2), we have

$$dd^c[\log |f|^2] = \delta_{\mathrm{div} f}.$$

so $-[\log |f|^2]$ is a Green current for $\mathrm{div} f$.

Consider the group $\widehat{Z}^p(X)$ of *arithmetic cycles*, i.e. pairs $(Z, g_Z)$ where $Z \in Z^p(X)$ and $g_Z$ is a Green current for $Z$, with addition defined componentwise. Let $\widehat{R}^p(X) \subset \widehat{Z}^p(X)$ be the subgroup generated by pairs $(\mathrm{div} f, -[\log |f|^2])$, where $f \in k(y)^*$ and $y \in X^{(p-1)}$, and by pairs $(0, \partial(u) + \overline{\partial}(v))$, where $u$ and $v$ are currents of type $(p-2, p-1)$ and $(p-1, p-2)$ respectively.

**Definition 1**  *The arithmetic Chow group of codimension $p$ of $X$ is the quotient $\widehat{CH}^p(X) = \widehat{Z}^p(X)/\widehat{R}^p(X)$.*

**1.2**    **Theorem 1**   *There are two exact sequences*

(1)   $CH^{p-1,p}(X) \xrightarrow{\rho} H^{p-1,p-1}(X) \xrightarrow{a} \widehat{CH}^p(X) \xrightarrow{(\zeta,\omega)} CH^p(X) \oplus Z^{p,p}(X)$
     $\xrightarrow{\mathrm{cl}} H^{p,p}(X) \longrightarrow 0,$

(2)   $CH^{p-1,p}(X) \xrightarrow{\rho} \widetilde{A}^{p-1,p-1}(X) \xrightarrow{a} \widehat{CH}^p(X) \xrightarrow{\zeta} CH^p(X) \longrightarrow 0.$

In this theorem the maps are defined as follows:

−    As we saw in §I.6.3, an element of $CH^{p-1,p}(X)$ is represented by $(f_y)$, $(f_y \in k(y)^*, \, y \in X^{(p-1)})$ such that $\sum_y \mathrm{div} f_y = 0$. The current $\sum_y -[\log|f_y|^2] \in D^{p-1,p-1}(X)$ satisfies

$$ dd^c\left(\sum_y -[\log|f_y|^2]\right) = -\delta_{\sum_y \mathrm{div} f_y} = 0 \,, $$

     hence defines an element in $H^{p-1,p-1}(X)$ and also in $\widetilde{A}^{p-1,p-1}(X)$, which we denote by $\rho((f_y))$. We will show in the proof of the Theorem that $\rho$ is well-defined on $CH^{p-1,p}(X)$.

−    Let $\mathrm{cl}(\eta) \in \widetilde{A}^{p-1,p-1}(X)$ denote the class of $\eta \in A^{p-1,p-1}(X)$; we put $a(\mathrm{cl}(\eta)) := [(0,[\eta])] \in \widehat{CH}^p(X)$, where 0 denotes the zero cycle of $Z^p(X)$ and $[(0,[\eta])]$ the class of $(0,[\eta])$. Similarly we define the map $a : H^{p-1,p-1}(X) \longrightarrow \widehat{CH}^p(X)$. The map $\zeta : \widehat{CH}^p(X)$ sends $[(Z, g_Z)]$ to $[Z]$ and $\omega : \widehat{CH}^p(X) \to Z^{p,p}(X)$ is given by $\omega([(Z, g_Z)]) = \omega_Z$, where $dd^c g_Z + \delta_Z = [\omega_Z]$. One checks easily that $a$, $\zeta$ and $\omega$ are well-defined.

−    The map $\mathrm{cl} : CH^p(X) \oplus Z^{p,p}(X) \longrightarrow H^{p,p}(X)$ is given by

$$ \mathrm{cl}([Z], \omega) = \mathrm{cl}(Z) - \mathrm{cl}(\omega), $$

     where $\mathrm{cl}(Z)$ denotes the class of the cycle $Z$ in $H^{p,p}(X)$   and $\mathrm{cl}(\omega)$ denotes the image of $\omega$ via the projection $Z^{p,p}(X) \to H^{p,p}(X)$.

*Proof of Theorem 1.*

(i) Obviously, cl and $\zeta$ are surjective maps.

(ii) To prove exactness at $CH^p(X) \oplus Z^{p,p}(X)$ note that $\mathrm{cl}([Z], \omega) = 0$ if and only if there exists $g \in D^{p-1,p-1}(X)$ with $dd^c g = [\omega] - \delta_Z$, i.e. $(\zeta, \omega)([(Z, g)]) = ([Z], \omega)$.

(iii) Let us prove the exactness at $\widehat{CH}^p(X)$ for (2) (resp. (1)). We have $\zeta([(Z, g_Z)]) = 0$ if and only if $Z = \sum_y \mathrm{div} f_y$ $(f_y \in k(y)^*, \, y \in X^{(p-1)})$, i.e.

$$ [(Z, g_Z)] = [(\sum_y \mathrm{div} f_y, g_Z)] = [(0, g_Z + \sum_y [\log|f_y|^2])] $$

$$ =: [(0, \widetilde{g})] \in \widehat{CH}^p(X). $$

But now $dd^c \widetilde{g} = [\omega_Z]$, and therefore $\widetilde{g} = [\eta] + \partial S_1 + \overline{\partial} S_2$ by Theorem II.1, with $\eta \in A^{p-1,p-1}(X)$, $S_1 \in D^{p-2,p-1}(X)$, $S_2 \in D^{p-1,p-2}(X)$. Hence we have $\zeta([(Z, g_Z)]) = 0$ if and only if

$$[(Z, g_Z)] = [(0, \widetilde{g})] = [(0, [\eta])] = a(\mathrm{cl}(\eta)),$$

where

$$\mathrm{cl}(\eta) \in \widetilde{A}^{p-1,p-1}(X).$$

To prove exactness at $\widehat{CH}^p(X)$ for (1), we note in addition that $dd^c\widetilde{g} = 0$, whence $\eta \in A^{p-1,p-1}(X)$ is a closed form and it defines an element $\mathrm{cl}(\eta) \in H^{p-1,p-1}(X)$. Again we have $(\zeta, \omega)([(Z, g_Z)]) = 0$ if and only if

$$[(Z, g_Z)] = [(0, \widetilde{g})] = [(0, [\eta])] = a(\mathrm{cl}(\eta)).$$

(iv) To prove exactness at $\widetilde{A}^{p-1,p-1}(X)$ (resp. $H^{p-1,p-1}(X)$) note that $a(\mathrm{cl}(\eta)) = 0$ if and only if

$$(0, [\eta]) = \sum_y (\mathrm{div} f_y, -[\log |f_y|^2]) + (0, \partial S_1 + \overline{\partial} S_2)$$

in $\widehat{Z}^p(X)$, where $f_y \in k(y)^*$, $y \in X^{(p-1)}$, $S_1$ is a current of type $(p-2, p-1)$, and $S_2$ a current of type $(p-1, p-2)$. This is equivalent to $\sum_y \mathrm{div} f_y = 0$ and

$$[\eta] = \sum_y -[\log |f_y|^2] + \partial S_1 + \overline{\partial} S_2 ,$$

in other words $\rho((f_y)) = \mathrm{cl}(\eta)$.

(v) We are left with proving that $\rho$ is well-defined. Recall that

$$CH^{p-1,p}(X) = E_{2X}^{p-1,-p}(X)$$

$$= \{(f_y) \in \bigoplus_{y \in X^{(p-1)}} k(y)^* : \sum_y \mathrm{div} f_y = 0\}/\mathrm{im} d_1,$$

where $d_1 : \bigoplus_{z \in X^{(p-2)}} K_2(k(z)) \longrightarrow \bigoplus_{y \in X^{(p-1)}} k(y)^*$ is given by the tame symbol (see §I.6.3), and hence we have to show that $\rho \circ d_1 = 0$. Now this is equivalent to the corresponding problem over $\mathbb{C}$, i.e., if $X$ is a smooth projective complex variety, $Z \subset X$ an irreducible subvariety of codimension $(p-2)$, and $f, g \in \mathbb{C}(Z)^*$, then $\rho \circ d_1(\{f, g\}) = 0$. To prove this, we may assume without loss of generality, that $\mathrm{div} f \cup \mathrm{div} g$ is a divisor with normal crossings, as the following remark shows. Let $\pi : \widetilde{Z} \to Z$ be a resolution of singularities of $\mathrm{div} f \cup \mathrm{div} g$ with $\pi$ proper and $D = \pi^{-1}(\mathrm{div} f \cup \mathrm{div} g)$ the divisor with normal crossings corresponding to $\mathrm{div} f \cup \mathrm{div} g$. By functoriality of $K$-theory, we have the commutative diagram

$$
\begin{array}{ccccc}
K_2(\mathbb{C}(\widetilde{Z})) & \xrightarrow{\ \widetilde{d}_1\ } & \bigoplus_{\widetilde{y}\in\widetilde{Z}^{(1)}}\mathbb{C}(\widetilde{y})^* & \xrightarrow{\ \rho\ } & D^{1,1}(\widetilde{Z})/(\mathrm{im}\partial+\mathrm{im}\overline{\partial}) \\
\Big\| & & \downarrow{\scriptstyle\pi_*} & & \downarrow{\scriptstyle\pi_*} \\
K_2(\mathbb{C}(Z)) & \xrightarrow{\ d_1\ } & \bigoplus_{y\in Z^{(1)}}\mathbb{C}(y)^* & \xrightarrow{\ \rho\ } & D^{1,1}(Z)/(\mathrm{im}\partial+\mathrm{im}\overline{\partial})
\end{array}
$$

and therefore $\rho\circ d_1=0$ will follow from $\rho\circ\widetilde{d}_1=0$, which is shown in the next lemma.

**Lemma 1**   *With the above notation, we have the formula*

$$
\rho\circ\widetilde{d}_1(\{f,g\})=\frac{1}{2i\pi}(\partial[\alpha]+\overline{\partial}[\beta]),
$$

*where $\alpha=\log|f|^2\wedge\overline{\partial}\,\log|g|^2$ and $\beta=\log|g|^2\wedge\partial\log|f|^2$.*

*Proof.*   Since the problem is local, we may assume that $\widetilde{Z}=\Delta^m=\{z\in\mathbb{C}^m:|z_j|<1\}$ and, by the linearity of the symbols $\{f,g\}$, we are reduced to the following two cases:

(a) $f=z_1$, $g=z_1$. Then the description of the map $d_1$ given in §I.6.3, implies

$$
\widetilde{d}_1(\{z_1,z_1\})=(-1)^{v(z_1)\cdot v(z_1)}\cdot z_1^{v(z_1)}\cdot z_1^{-v(z_1)}=-1,
$$

hence $\rho\circ\widetilde{d}_1(\{z_1,z_1\})=-[\log|-1|^2]=0$. On the other hand, since the form $\partial\alpha+\overline{\partial}\beta$ vanishes outside the origin, if $\omega\in A^{m-1,m-1}(\Delta^m)$ we get

$$
(\partial[\alpha]+\overline{\partial}[\beta])(\omega)=\lim_{\epsilon\to0}\int_{|z_1|=\epsilon}(\alpha+\beta)\wedge\omega\ .
$$

Using polar coordinates $z_1=re^{i\vartheta}$, $\alpha+\beta$ becomes $(2/r)\cdot\log(r^2)dr$, and this integral vanishes as $\epsilon$ goes to zero.

(b) $\mathrm{div}f$ and $\mathrm{div}g$ intersect properly. Again the description of the map $d_1$ gives

$$
\rho\circ\widetilde{d}_1(\{f,g\})=[\log|f|^2]\wedge\delta_{\mathrm{div}g}-[\log|g|^2]\wedge\delta_{\mathrm{div}f}.
$$

By definition of the $*$-product (see Definition II.4), we have

$$
\rho\circ\widetilde{d}_1(\{f,g\})=-[\log|f|^2]*[\log|g|^2]+[\log|g|^2]*[\log|f|^2].
$$

The result now follows from the following fact. If $Y$ and $Z$ intersect properly, and $g_Y$ (resp. $g_Z$) is a Green form of logarithmic type for $Y$ (resp. $Z$), we have

$$
[g_Y]*[g_Z]-[g_Z]*[g_Y]=1/2i\pi\cdot(\partial[g_Y\wedge\overline{\partial}g_Z]+\overline{\partial}[g_Z\wedge\partial g_Y]).
$$

To prove this identity we use the definition of the $*$-product to obtain

$$[g_Y] * [g_Z] - [g_Z] * [g_Y]$$
$$= [g_Y] \wedge \delta_Z + [\omega_Y] \wedge [g_Z] - [g_Z] \wedge \delta_Y - [\omega_Z] \wedge [g_Y]$$
$$= -[g_Y] \wedge ([\omega_Z] - \delta_Z) + [g_Z] \wedge ([\omega_Y] - \delta_Y)$$
$$= -[g_Y] \wedge dd^c[g_Z] + [g_Z] \wedge dd^c[g_Y]$$
$$= 1/(2i\pi) \cdot (\partial([g_Y] \wedge \bar\partial[g_Z]) + \bar\partial([g_Z] \wedge \partial[g_Y]))$$
$$= 1/(2i\pi) \cdot (\partial[g_Y \wedge \bar\partial g_Z] + \bar\partial[g_Z \wedge \partial g_Y]),$$

where the last equality can be justified for forms of logarithmic type (see [GS2]) .                                                                          □

**1.3**    It has been conjectured (Bass; see also [Ger], Problem 21, for higher $K$-groups) that $CH^p(X)$ is a finitely generated $\mathbb{Z}$-module. This is known to be true when $p$ is equal to $0, 1$ or the dimension of $X$.

## 2. The intersection pairing

**2.1**    Recall that $X$ is a regular scheme, projective and flat over $\mathbb{Z}$. We denote by $X_{\mathbb{Q}}$ its generic fibre $X \times_{\mathrm{Spec}\mathbb{Z}} \mathrm{Spec}\mathbb{Q}$. Let us introduce the following groups

$$Z^p_{\mathrm{fin}}(X) := \{Z \in Z^p(X) : |Z| \cap X_{\mathbb{Q}} = \emptyset\},$$

where $|Z|$ is the support of the cycle $Z$,

$$CH^p_{fin}(X) := Z^p_{\mathrm{fin}}(X)/\langle \mathrm{div} f \rangle,$$

where $y \in X^{(p-1)} - X_{\mathbb{Q}}$ and $f \in k(y)^*$, and

$$\widehat{Z}^p(X_{\mathbb{Q}}) := \{(Z, g_Z) : Z \in Z^p(X_{\mathbb{Q}}), g_Z \text{ a Green current for } Z\}.$$

Note that any cycle $Z \in Z^p(X)$ can be written uniquely $Z = Z_1 + Z_2$ with $Z_1 \in Z^p_{\mathrm{fin}}(X)$ and $Z_2 \in Z^p(X_{\mathbb{Q}})$.

**Lemma 2**    *We have an isomorphism*

$$\widehat{CH}^p(X) \cong (CH^p_{\mathrm{fin}}(X) \oplus \widehat{Z}^p(X_{\mathbb{Q}}))/\langle \widehat{\mathrm{div}}(f); (0, 0, \mathrm{im}\partial + \mathrm{im}\bar\partial) \rangle,$$

*where* $y \in X^{(p-1)}_{\mathbb{Q}}$, $f \in k(y)^*$ *and* $\widehat{\mathrm{div}}(f) = (\mathrm{div} f, -[\log|f|^2])$.

*Proof.*    There is a natural map

$$\widehat{CH}^p(X) \longrightarrow (CH^p_{\mathrm{fin}}(X) \oplus \widehat{Z}^p(X_{\mathbb{Q}}))/\langle \widehat{\mathrm{div}}(f); (0, 0, \mathrm{im}\partial + \mathrm{im}\bar\partial) \rangle,$$

which is obviously surjective. To prove injectivity, one simply notes that for $f \in k(y)^*$, $y \in X^{(p-1)} - X_{\mathbb{Q}}$, the current $[\log|f|^2]$ is zero.

**Remark**    Let $Y \subset X$ be a closed subscheme such that $\mathrm{codim}_{X_{\mathbb{Q}}}(Y_{\mathbb{Q}}) = p$. Then the natural map

$$Z_Y^p(X) \longrightarrow Z_{\mathrm{fin}}^p(X) \oplus Z_{Y_{\mathbb{Q}}}^p(X_{\mathbb{Q}})$$

induces a map

$$CH_Y^p(X) \longrightarrow CH_{\mathrm{fin}}^p(X) \oplus Z_{Y_{\mathbb{Q}}}^p(X_{\mathbb{Q}})$$

and $Z_{Y_{\mathbb{Q}}}^p(X_{\mathbb{Q}}) = CH_{Y_{\mathbb{Q}}}^p(X_{\mathbb{Q}})$.

## 2.2

**Theorem 2**    *There is a pairing*

$$\widehat{CH}^p(X) \otimes \widehat{CH}^q(X) \longrightarrow \widehat{CH}^{p+q}(X)_{\mathbb{Q}}$$

*with the following properties:*

(i)    $\bigoplus_{p \geq 0} \widehat{CH}^p(X)_{\mathbb{Q}}$ *is a commutative graded unitary $\mathbb{Q}$-algebra.*

(ii)    *The map*

$$(\zeta, \omega) : \bigoplus_{p \geq 0} \widehat{CH}^p(X)_{\mathbb{Q}} \longrightarrow \bigoplus_{p \geq 0} (CH^p(X) \oplus Z^{p,p}(X))_{\mathbb{Q}}$$

*(cf. Theorem 1) is a $\mathbb{Q}$-algebra homomorphism.*

*Proof.*    Let $[(Y, g_Y)] \in \widehat{CH}^p(X)$ and $[(Z, g_Z)] \in \widehat{CH}^q(X)$. To define the pairing, we may assume that $Y, Z$ are irreducible and then extend the definition by linearity. Furthermore, we assume first that $Y_{\mathbb{Q}}$ and $Z_{\mathbb{Q}}$ intersect properly, i.e., $\mathrm{codim}_{X_{\mathbb{Q}}}(Y_{\mathbb{Q}} \cap Z_{\mathbb{Q}}) = p + q$ . We have $[Y] \in CH_Y^p(X)$, $[Z] \in CH_Z^q(X)$, hence by Theorem I.2, $[Y] \cdot [Z] \in CH_{Y \cap Z}^{p+q}(X)_{\mathbb{Q}}$ and we again denote by $[Y] \cdot [Z]$ the image of this element by the map

$$CH_{Y \cap Z}^{p+q}(X)_{\mathbb{Q}} \longrightarrow CH_{\mathrm{fin}}^{p+q}(X)_{\mathbb{Q}} \oplus Z_{Y_{\mathbb{Q}} \cap Z_{\mathbb{Q}}}^{p+q}(X_{\mathbb{Q}})_{\mathbb{Q}}$$

considered in the remark in §2.1. We then define

$$[(Y, g_Y)] \cdot [(Z, g_Z)] := [([Y] \cdot [Z], g_Y * g_Z)]$$

in

$$CH_{\mathrm{fin}}^{p+q}(X)_{\mathbb{Q}} \oplus \widehat{Z}^{p+q}(X_{\mathbb{Q}})_{\mathbb{Q}}/\langle \widehat{\mathrm{div}}(f); (0, 0, \mathrm{im}\partial + \mathrm{im}\overline{\partial}) \rangle_{\mathbb{Q}} \cong \widehat{CH}^{p+q}(X)_{\mathbb{Q}}$$

(see Lemma 2). Note that the $*$-product is well defined, since $Y(\mathbb{C})$ and $Z(\mathbb{C})$ intersect properly ; see §II.3.1.

We now turn to the case where $Y_{\mathbb{Q}}$ and $Z_{\mathbb{Q}}$ do *not* intersect properly. By the Moving Lemma over $\mathbb{Q}$ [RJ], we know that there are rational functions $f_y \in k(y)^*$, $y \in X_{\mathbb{Q}}^{(p-1)}$ such that $Y + \sum_y \mathrm{div} f_y$ and $Z$ intersect properly on $X_{\mathbb{Q}}$, and we are reduced to the proper intersection case treated above. It remains to show that, if $g_y \in k(y)^*$, $y \in X_{\mathbb{Q}}^{(p-1)}$ is another choice of rational functions such that $(Y + \sum_y \mathrm{div}(g_y))_{\mathbb{Q}}$ and $Z_{\mathbb{Q}}$ intersect properly, then the cycle

$$\left(\sum_y \widehat{\mathrm{div}}(f_y) - \sum_y \widehat{\mathrm{div}}(g_y)\right) \cdot (Z, g_Z)$$

lies in the subgroup $\widehat{R}^{p+q}(X)_{\mathbb{Q}}$ of $\widehat{Z}^{p+q}(X)_{\mathbb{Q}}$ (see §1.1). To prove this, we need a Moving Lemma for $K_1$-chains, which we state without proof (see [GS2]), using the notation of §I.6.3.

**Lemma 3**   *Let*

$$(h_y) \in \bigoplus_{y \in X_{\mathbb{Q}}^{(p-1)}} k(y)^* = \bigoplus_{y \in X_{\mathbb{Q}}^{(p-1)}} K_1(k(y)) = E_1^{p-1,-p}(X_{\mathbb{Q}})$$

*be a $K_1$-chain on $X_{\mathbb{Q}}$ such that $(\sum_y \mathrm{div}(h_y))_{\mathbb{Q}}$ and $Z_{\mathbb{Q}}$ intersect properly. Then there exists $u \in \bigoplus_{z \in X_{\mathbb{Q}}^{(p-2)}} K_2(k(z))$ such that, if $(\widetilde{h}_y) := (h_y) + d_1(u)$, then $(\mathrm{div}(\widetilde{h}_y))_{\mathbb{Q}}$ intersects $Z_{\mathbb{Q}}$ properly for all $y \in X_{\mathbb{Q}}^{(p-1)}$.*

In particular, Lemma 3 gives

$$\sum_y \mathrm{div}(h_y) = \sum_y \mathrm{div}(\widetilde{h}_y) \ ,$$

because $\mathrm{div} \circ d_1 = d_1^2 = 0$, and $\sum_y [\log|h_y|^2] - \sum_y [\log|\widetilde{h}_y|^2]$ lies in $(\mathrm{im}\partial + \mathrm{im}\overline{\partial})$, since $\rho \circ d_1 = 0$, as shown in the proof of Theorem 1. So, given $f_y$ and $g_y$ as above, we can find $(\widetilde{h}_y) \in \bigoplus_{y \in X_{\mathbb{Q}}^{(p-1)}} k(y)^*$ such that $\mathrm{div}(\widetilde{h}_y)_{\mathbb{Q}}$ intersects $Z_{\mathbb{Q}}$ properly for all $y \in X_{\mathbb{Q}}^{(p-1)}$ and

$$\sum_y (\widehat{\mathrm{div}}(f_y) - \widehat{\mathrm{div}}(g_y) - \widehat{\mathrm{div}}(\widetilde{h}_y)) \cdot (Z, g_Z)$$

lies in $(0, \mathrm{im}\partial + \mathrm{im}\overline{\partial})$. We are finished if we show that each summand $\widehat{\mathrm{div}}(\widetilde{h}_y) \cdot (Z, g_Z)$ lies in $\widehat{R}^{p+q}(X)_{\mathbb{Q}}$. To simplify the notation, we write $h$ for $\widetilde{h}_y$, and furthermore we assume that $Z_{\mathrm{fin}} = \emptyset$.

Let $W$ be the support of $h$. Then $|W_{\mathbb{Q}}| \cap |Z_{\mathbb{Q}}| = S \cup T$, where $\mathrm{codim}_{X_{\mathbb{Q}}} S = p + q - 1$ and $\mathrm{codim}_{X_{\mathbb{Q}}} T < p + q - 1$. In $CH_{S \cup T}^{p+q-1}(X_{\mathbb{Q}})_{\mathbb{Q}}$ we have

$$[W_{\mathbb{Q}}] \cdot [Z_{\mathbb{Q}}] = \sum_i \mu_i [S_i] + \tau,$$

where $S_i$ are the irreducible components of $S$, $[S_i] \in CH_S^{p+q-1}(X_{\mathbb{Q}})_{\mathbb{Q}}$, $\tau \in CH_T^{p+q-1}(X_{\mathbb{Q}})_{\mathbb{Q}}$, and $\mu_i = \mu_i(W_{\mathbb{Q}}, Z_{\mathbb{Q}})$ are Serre's intersection multiplicities. Since $(\mathrm{div}(h))_{\mathbb{Q}}$ intersects $Z_{\mathbb{Q}}$ properly, the restriction $h|_{S_i}$ of $h$ to $S_i$ is not identically zero (otherwise $(\mathrm{div}(h))_{\mathbb{Q}}$ would have a component of codimension $p + q - 1$, which is impossible), and $h|_T$ is a unit, since $(\mathrm{div}(h))_{\mathbb{Q}} \cap T = \emptyset$.

We now define the product of the $K_1$-chain

$$h \in \bigoplus_{y \in X_{\mathbb{Q}}^{(p-1)}} k(y)^* = \bigoplus_{y \in X_{\mathbb{Q}}^{(p-1)}} K_1(k(y))$$

with the codimension $q$-cycle

$$Z \in \bigoplus_{x \in X_{\mathbb{Q}}^{(q)}} \mathbb{Z} = \bigoplus_{x \in X_{\mathbb{Q}}^{(q)}} K_0(k(x))$$

to be the $K_1$-chain

$$h \cdot Z := \prod_i (h|_{S_i})^{\mu_i} \cdot (h|_T \cdot t) \in \bigoplus_{y \in X_{\mathbb{Q}}^{(p+q-1)}} K_1(k(y)),$$

where $t \in Z_T^{p+q-1}(X_{\mathbb{Q}})_{\mathbb{Q}}$ is a representative of $\tau$; the product $(h|_T \cdot t)$ has to be understood in $K$-theoretic terms. Note that $h \cdot Z$ is only defined up to $\mathrm{im}\, d_1$, so $\widehat{\mathrm{div}}(h \cdot Z)$ is well-defined, because $\mathrm{div} \circ d_1 = 0$ and $\rho \circ d_1 = 0$. The proof of Theorem 2 will be complete if we show that $\widehat{\mathrm{div}} h \cdot (Z, g_Z) - \widehat{\mathrm{div}}(h \cdot Z)$ lies in $(0, \mathrm{im}\, \partial + \mathrm{im}\, \bar{\partial})$.

To prove this, let $H \in k(X)^*$ be such that $H|_W = h$. Since $\mathrm{div} H \cdot W = \mathrm{div}(H|_W) = \mathrm{div} h$ by [Fu], Chapter 2, we get

$$\mathrm{div} h \cdot Z = (\mathrm{div} H \cdot W) \cdot Z_{\mathbb{Q}} = \mathrm{div} H \cdot (W_{\mathbb{Q}} \cdot Z_{\mathbb{Q}})$$

$$= \mathrm{div}(H) \cdot \left( \sum_i \mu_i S_i + t \right)$$

$$= \sum_i \mu_i \mathrm{div}(H|_{S_i}) + \mathrm{div}(H|_t)$$

$$= \sum_i \mu_i \mathrm{div}(h|_{S_i}) + \mathrm{div}(h|_T \cdot t) = \mathrm{div}(h \cdot Z).$$

The Green current component of $\widehat{\mathrm{div}} h \cdot (Z, g_Z)$, because of commutativity and the definition of the $*$-product, is equal to

$$(-[\log |h|^2]) * g_Z = g_Z * (-[\log |h|^2]) = g_Z \wedge \delta_{\mathrm{div} h} + [\omega_Z] \wedge (-[\log |h|^2])$$

modulo $(\mathrm{im}\, \partial + \mathrm{im}\, \bar{\partial})$. For the Green current of $\widehat{\mathrm{div}}(h \cdot Z)$, using Theorem II.4 (or rather a refinement of it for improper intersections), the associativity, the commutativity and the definition of the $*$-product, we get

$$-[\log |h \cdot Z|^2] = -[\log |H|^2] \wedge \delta_{W \cdot Z}$$

$$= -[\log |H|^2] * (g_W * g_Z) = g_Z * (-[\log |H|^2] * g_W)$$

$$= g_Z \wedge \delta_{\mathrm{div} H \cdot W} + [\omega_Z] \wedge (-[\log |H|^2] * g_W)$$

$$= g_Z \wedge \delta_{\mathrm{div} h} - [\omega_Z] \wedge [\log |H|^2] \wedge \delta_W$$

$$= g_Z \wedge \delta_{\mathrm{div} h} + [\omega_Z] \wedge (-[\log |h|^2])$$

modulo $(\mathrm{im}\, \partial + \mathrm{im}\, \bar{\partial})$. We refer to [GS2] for the justification of these equalities.

The above construction turns $\bigoplus_{p \geq 0} \widehat{CH}^p(X)_{\mathbb{Q}}$ into a commutative graded $\mathbb{Q}$-algebra with unit $[(X, 0)] \in \widehat{CH}^{\dim X}(X)$; commutativity and associativity follow from the corresponding properties of the intersection product on regular schemes (see Chapter I) and the $*$-product (see Chapter II). The multiplicativity of the map $(\zeta, \omega)$ is also clear, since, by Theorem II.4, we have

$$(\zeta, \omega)([(Y, g_Y)] \cdot [(Z, g_Z)]) = (\zeta, \omega)([([Y] \cdot [Z], g_Y * g_Z)])$$
$$= ([Y] \cdot [Z], [\omega_Y \wedge \omega_Z]).$$

## 2.3
## Remarks
**2.3.1** From the definitions we get the following useful formula:

$$(3) \qquad a(\eta)x = a(\eta\omega(x))$$

for any $x \in \widehat{CH}^q(X)$ and $\eta \in \tilde{A}^{p-1, p-1}(X)$.

**2.3.2** If $q = 1$ in Theorem 2, we do not have to tensor with $\mathbb{Q}$, i.e. we get a pairing

$$(4) \qquad \widehat{CH}^p(X) \otimes \widehat{CH}^1(X) \longrightarrow \widehat{CH}^{p+1}(X),$$

because in this situation there is a Moving Lemma. Namely, if $Y \in Z^p(X)$ and $Z \in Z^1(X)$, there exists $Z'$ rationally equivalent to $Z$ such that $Z'$ intersects $Y$ properly.

Indeed, let $Y_i$, $i \in I$, be the irreducible components of $Y$ and $y_i$ the generic point of $Y_i$. The semi-local ring of functions on $X$ regular near all $y_i$, $i \in I$, is a unique factorization domain since it is regular. So, in the spectrum of this ring, $Z$ is the divisor of some rational function. In other words, there exists $Z'$ rationally equivalent to $Z$ on $X$ which does not contain any of the $Y_i$'s.

Notice also that Green currents in codimension one are easy to construct — they all come from the Poincaré–Lelong formula, see Proposition 1 below. Therefore the pairing (4) is easier to define than the general pairing in Theorem 2. It turns out that, for many applications, this pairing is sufficient; see [F4].

**2.3.3** Assume that $X$ is smooth on the ring of integers in a number field. Then we can use Fulton's approach to intersection theory [Fu], so we do not need to tensor with $\mathbb{Q}$ in Theorem 2 (see [GS2]).

**2.3.4** One may wonder whether there exist higher arithmetic $K$-groups $\widehat{K}_m$ (see Definition IV.5 below for the case $m = 0$) and if a formula à la Gersten $\widehat{CH}^p(X) = H^p(X, \widehat{K}_p)$ would hold, similar to §I.6.5.

One possible definition when $m > 0$ is

$$\widehat{K}_m(X) = \pi_{m+1} \text{ (homotopy fiber of the regulator map)},$$

where the regulator map is viewed as a map from $BQP(X)$ to a product of Eilenberg–Maclane spaces; see [Be] and [G2]. This definition provides an exact sequence

(5)
$$\cdots \xrightarrow{\rho} \bigoplus_{m+1+k=2i} H^k_D(X, \mathbb{R}(i)) \xrightarrow{a} \widehat{K}_m(X) \longrightarrow K_m(X)$$
$$\xrightarrow{\rho} \bigoplus_{m+k=2i} H^k_D(X, \mathbb{R}(i)) \longrightarrow \cdots,$$

where $\rho$ is the Beilinson regulator with values in Deligne cohomology [Be]. If $\widehat{\mathcal{K}}_p$ is the Zariski sheaf attached to the presheaf $U \mapsto \widehat{K}_p(U)$ defined by (5), a guess is that $H^p(X, \widehat{\mathcal{K}}_p)$ is isomorphic to the subgroup $\ker(\omega)$ of $\widehat{CH}^p(X)$ (see [S2]).

By means of a generalization of the Bott-Chern character classes, X. Wang has defined an explicit simplicial set whose homotopy groups should be the groups $\widehat{K}_m(X)$ [Wa].

## 3. Functoriality

### 3.1
**Theorem 3**  *Let $X, Y$ be regular schemes, projective and flat over $\mathbb{Z}$ and let $f : Y \to X$ be a morphism.*

(i)    *There is a pull-back homomorphism $f^* : \widehat{CH}^p(X) \to \widehat{CH}^p(Y)_{\mathbb{Q}}$. It is multiplicative, i.e., given $\alpha \in \widehat{CH}^p(X)$ and $\beta \in \widehat{CH}^q(X)$, one has*

$$f^*(\alpha \cdot \beta) = f^*(\alpha) \cdot f^*(\beta).$$

(ii)   *If $f$ is proper, $f_{\mathbb{Q}} : Y_{\mathbb{Q}} \to X_{\mathbb{Q}}$ is smooth and $X, Y$ are equidimensional, then there is a push-forward homomorphism*

$$f_* : \widehat{CH}^p(Y) \to \widehat{CH}^{p-\delta}(X) \quad (\delta := \dim Y - \dim X).$$

(iii)  *The following projection formula holds:*

$$f_*(f^*(\alpha) \cdot \beta) = \alpha \cdot f_*(\beta) \in \widehat{CH}^{p+q-\delta}(X)_{\mathbb{Q}}$$

*if $\alpha \in \widehat{CH}^p(X)$ and $\beta \in \widehat{CH}^q(Y)$.*

(iv)   *Given two maps $f : Y \to X$ and $g : Z \to Y$ one has $(fg)^* = g^* f^*$ and $(fg)_* = f_* g_*$.*

*Proof.*
(i) Let $[(Z, g_Z)] \in \widehat{CH}^p(X)$. To define the pull-back, we may assume that

$Z$ is irreducible and then extend this definition by linearity. Furthermore, we may assume first that $\mathrm{codim}_{Y_\mathbb{Q}}(f^{-1}(Z)_\mathbb{Q}) = p$, and reduce afterwards the general case to this situation. Since $K$-theory is contravariant, we may apply Theorem I.3, to get a class $f^*[Z] \in CH^p_{f^{-1}(Z)}(Y)_\mathbb{Q}$. We denote also by $f^*[Z]$ its image by the map

$$CH^p_{f^{-1}(Z)}(Y)_\mathbb{Q} \longrightarrow CH^p_{fin}(Y)_\mathbb{Q} \oplus Z^p_{f^{-1}(Z)_\mathbb{Q}}(Y_\mathbb{Q})_\mathbb{Q}$$

(see the remark in §2.1). Furthermore, by §II.3.2, the pull-back $f^*g_Z$ of the Green current $g_Z$ can be defined. We put

$$f^*[(Z, g_Z)] := [(f^*[Z], f^*g_Z)] \in \widehat{CH}^p(Y)_\mathbb{Q}.$$

As in the proof of Theorem 2, we can remove the assumption

$$\mathrm{codim}_{Y_\mathbb{Q}}(f^{-1}(Z)_\mathbb{Q}) = p$$

by noting that there exist rational functions $f_y \in k(y)^*$, $y \in Y_\mathbb{Q}^{(p-1)}$, such that $\mathrm{codim}_{Y_\mathbb{Q}}((f^{-1}(Z) + \sum_y \mathrm{div} f_y)_\mathbb{Q}) = p$, and one is reduced to the above definition. Of course, we are left with proving that this definition of $f^*[(Z, g_Z)]$ does not depend on the choice of the rational functions $f_y$. But this can be done in a way similar to the proof of Theorem 2.

We shall not prove that $f^*$ is a ring homomorphism; see [GS2].

(ii) We define first a map from $\widehat{Z}^p(Y)$ to $\widehat{Z}^{p-\delta}(X)$ as follows. Let $(Z, g_Z) \in \widehat{Z}^p(Y)$ with $Z$ irreducible, i.e., $Z = \overline{\{z\}}$, where $z$ is the generic point of $Z$. Following [Fu], 1.4, we put

$$f_*(Z) := \begin{cases} [k(z) : k(f(z))] \cdot \overline{\{f(z)\}} & \text{if } \dim f(z) = \dim z; \\ 0 & \text{if } \dim f(z) < \dim z. \end{cases}$$

Next we observe that, given a differential $\eta$ on $X(\mathbf{C})$ (of the appropriate degree), we have

$$(f_*\delta_Z)(\eta) = \delta_Z(f^*\eta) = \int_{Z(\mathbf{C})} f^*\eta = \int_{Z(\mathbf{C})} f^*(\eta|_{f(Z(\mathbf{C}))})$$

$$= \begin{cases} \deg(Z(\mathbf{C})/f(Z(\mathbf{C}))) \cdot \int_{f(Z(\mathbf{C}))} \eta & \text{if } \dim f(Z(\mathbf{C})) = \dim Z(\mathbf{C}); \\ 0 & \text{if } \dim f(Z(\mathbf{C})) < \dim Z(\mathbf{C}). \end{cases}$$

Hence we have $f_*\delta_Z = \delta_{f_*(Z)}$, from which we deduce

$$dd^c(f_*g_Z) = [f_*\omega_Z] - \delta_{f_*(Z)}.$$

Here $f_*\omega_Z \in A^{p-\delta, p-\delta}(X)$ is obtained by integration along the fibers, because $f_\mathbb{Q} : Y_\mathbb{Q} \to X_\mathbb{Q}$ is smooth. Therefore $f_*g_Z$ defines a Green current for $f_*(Z)$, and we put

$$f_*(Z, g_Z) := (f_*Z, f_*g_Z) \in \widehat{Z}^{p-\delta}(X).$$

We are left with checking that this definition induces a map

$$f_* : \widehat{CH}^p(Y) \to \widehat{CH}^{p-\delta}(X).$$

For that purpose, let $h \in k(W)^*$, $W = \overline{\{y\}}$, $y \in Y^{(p-1)}$ and $f_W := f|_W :$ $W \to W' := f(W)$. By [Fu], Proposition 1.4, we then have

$$f_*(\text{div}(h)) = \begin{cases} \text{div}(\text{Norm}_{k(W)/k(W')}(h)) & \text{if } \dim W' = \dim W; \\ 0 & \text{if } \dim W' < \dim W. \end{cases}$$

In the former case, the map $W(\mathbb{C}) \longrightarrow W'(\mathbb{C})$ is finite, hence there is an open set $U \subset W'(\mathbb{C})$ such that $f_W : f_W^{-1}(U) \to U$ is finite and étale. Hence for any $L^1$-function $\varphi$ on $W(\mathbb{C})$ and $u \in U$, we have

$$f_{W*}(\varphi)(u) = \sum_{w \in f_W^{-1}(u)} \varphi(w).$$

From this we deduce

$$f_*[\log |h|^2] = [\log |\text{Norm}_{k(W(\mathbb{C}))/k(W'(\mathbb{C}))}(h)|^2],$$

which completes the proof of (ii).

(iii) Let $\alpha = [(Z, g_Z)] \in \widehat{CH}^p(X)$ and $\beta = [(W, g_W)] \in \widehat{CH}^q(Y)$. By §I.3.2 and Theorem 3, we know the validity of the projection formula for cycle-classes. Therefore, we are left to prove it for Green currents. By definition of the $*$-product, we have

$$f_*(f^* g_Z * g_W) = f_*(f^* g_Z \wedge \delta_W + [f^* \omega_Z] \wedge g_W) = h_*(h^* g_Z) + [\omega_Z] \wedge f_* g_W,$$

where $h : W \longrightarrow X$ denotes the map induced by $f$. But as in part (ii) of the proof, we have $h_*(h^* g_Z) = g_Z \wedge \delta_{f_*(W)}$, whence

$$f_*(f^* g_Z * g_W) = g_Z \wedge \delta_{f_*(W)} + [\omega_Z] \wedge f_* g_W = g_Z * f_* g_W,$$

as claimed.

We refer to [GS2] for the proof of (iv).          $\square$

**Remark**  Because $Y$ can be embedded into a smooth variety over $\mathbb{Z}$ (for instance a projective space), one can use Fulton's approach to intersection theory [Fu] to avoid tensoring with $\mathbb{Q}$ when defining $f^*$ in Theorem 3; see [GS2].

## 4. Examples

**4.1**    Let $X$ be a regular scheme, projective and flat over $\mathbb{Z}$. Then

$$\widehat{CH}^0(X) = CH^0(X) \cong \mathbb{Z}^{\pi_0(X)}.$$

**4.2**    A line bundle on $X$ with a smooth hermitian metric invariant under $F_\infty$ on the complex line bundle induced by $L$ on $X(\mathbb{C})$ is called an *hermitian line bundle*. Denote by $\widehat{\mathrm{Pic}}(X)$ the group of isomorphism classes of hermitian line bundles on $X$, where the group structure is given by tensor product and isomorphisms are those algebraic isomorphisms on $X$ which preserve the metrics.

**Proposition 1**    [D] *There is an isomorphism*

$$\widehat{c}_1 : \widehat{\mathrm{Pic}}(X) \longrightarrow \widehat{CH}^1(X)$$

*mapping the class of* $(L, \|\cdot\|)$ *to the class of* $[(\mathrm{div}\, s, -[\log \|s\|^2])]$, *for any rational section* $s$ *of* $L$.

*Proof.*    The class $\widehat{c}_1(L, \|\cdot\|)$ does not depend on the choice of $s$ since any other rational section can be written $s' = fs$, where $f$ is a rational function on $X$. The inverse of the map $\widehat{c}_1$ is obtained by sending $(Z, g_Z)$ to the isomorphism class of $(\mathcal{O}_X(Z), \|\cdot\|)$, where the metric $\|\cdot\|$ is locally given by the formula $\|f\|^2 = |f|^2 \mathrm{e}^{-g_Z}$; note that this defines a smooth metric, because the function $g_Z - \log|f|^2$ is smooth.                              □

**4.3**    Let $X = \mathrm{Spec}\,\mathcal{O}_F$, where $\mathcal{O}_F$ denotes the ring of integers of a number field $F$. In this special situation we have

$$CH^{0,1}(X) = \mathcal{O}_F^* \quad (\text{see } \S I.6.3),$$

$$\widetilde{A}^{0,0}(X) = A^{0,0}(X) = \bigoplus_{\sigma \in \Sigma} \mathbb{R};$$

$\Sigma$ denoting a set of $r_1$ real and $r_2$ non-conjugate complex imbeddings of $F$,

$$CH^1(X) = \mathrm{cl}(\mathcal{O}_F) \quad (\text{the ideal class group of } \mathcal{O}_F).$$

Theorem 1 (2) then gives the exact sequence

(6)        $\mathcal{O}_F^* \xrightarrow{\rho} \mathbb{R}^{r_1+r_2} \xrightarrow{a} \widehat{\mathrm{Pic}}(X) \xrightarrow{\varsigma} \mathrm{cl}(\mathcal{O}_F) \longrightarrow 0.$

Notice that $\rho$ is, up to a factor $-2$, the classical Dirichlet regulator map, hence $\ker \rho = \mu_F$, the roots of unity of $F$.

There is also a *degree map*

$$\widehat{\deg} : \widehat{\mathrm{Pic}}(X) \longrightarrow \mathbb{R}$$

mapping $(L, \|\cdot\|)$ to $\log([L : (\mathcal{O}_F s)]) - \log(\prod_\sigma \|s\|_\sigma)$ for any $s \in L\backslash\{0\}$, $\sigma$ running over all complex imbeddings of $F$. In particular one gets an isomorphism

(7)                         $\widehat{\deg} : \widehat{\mathrm{Pic}}(\mathbb{Z}) \cong \mathbb{R},$

by sending $(L, \| \cdot \|)$ to $- \log \|s\|$, where $s$ is a generator of the rank one $\mathbf{Z}$-module $L$.

Endow $\widehat{\mathrm{Pic}}(X)$ with the quotient topology in each fiber of $\zeta$. Then the compactness of $\widehat{\mathrm{Pic}}^0(X) := \ker(\widehat{\deg})$ is equivalent to the finiteness of the ideal class group $\mathrm{cl}(\mathcal{O}_F)$ and the Dirichlet Unit Theorem (i.e. $\rho(\mathcal{O}_F^*)$ is a lattice of rank $r_1 + r_2 - 1$ in $\mathbb{R}^{r_1+r_2}$). Furthermore, we note that $\mathrm{vol}(\widehat{\mathrm{Pic}}^0(X)) = h_F \cdot R_F$, where $h_F$ (resp. $R_F$) denotes the class number (resp. regulator) of $F$. See [W2] and [Sz] for more details on this.

## 5. Arakelov varieties

**5.1**    Let $X$ be a regular scheme, projective and flat over $\mathbf{Z}$ and $\omega_0$ be a Kähler metric on $X(\mathbb{C})$, invariant under $F_\infty$.

**Definition 2**   *The pair $\overline{X} := (X, \omega_0)$ is called an Arakelov variety.*

By the Hodge decomposition (see Theorem II.1), we have
$$A^{p,p}(X) = \mathcal{H}^{p,p}(X) \oplus \mathrm{im}\, d \oplus \mathrm{im}\, d^*,$$
where $\mathcal{H}^{p,p}(X) = \ker \Delta \subset A^{p,p}(X)$ denotes the space of real harmonic forms on $X(\mathbb{C})$ of type $(p,p)$, invariant under $F_\infty$ up to the factor $(-1)^p$; we keep the notation of §1.

**Definition 3**   *The group $CH^p(\overline{X}) := \omega^{-1}(\mathcal{H}^{p,p}(X)) \subset \widehat{CH}^p(X)$ is called the p-th Arakelov Chow group of $X$.*

There is also a Hodge decomposition for currents
$$D^{p,p}(X) = \mathcal{H}^{p,p}(X) \oplus \mathrm{im}\, d \oplus \mathrm{im}\, d^*,$$
and we denote by $H : D^{p,p}(X) \longrightarrow \mathcal{H}^{p,p}(X)$ the orthogonal projection (harmonic projection).

**Proposition 2**   *We have*

(i)   $CH^p(\overline{X}) \cong (Z^p(X) \oplus \mathcal{H}^{p-1,p-1}(X))/\langle \mathrm{div} f, -H[\log |f|^2]\rangle$
     *(with $f \in k(y)^*$, $y \in X^{(p-1)}$).*

(ii)   $CH^p(\overline{X})$ *is a direct summand of $\widehat{CH}^p(X)$.*

(iii)   *There is an exact sequence*

$$CH^{p-1,p}(X) \xrightarrow{\rho} H^{p-1,p-1}(X) \xrightarrow{a} CH^p(\overline{X}) \xrightarrow{\zeta} CH^p(X) \longrightarrow 0.$$

*Proof.*

(i) Let $Z^p(\overline{X}) = \{(Z, g_Z) : Z \in Z^p(X), \omega_Z \in \mathcal{H}^{p,p}(X)\}$. There is an isomorphism $Z^p(\overline{X}) \cong Z^p(X) \oplus \mathcal{H}^{p-1,p-1}(X)$ sending $(Z, g_Z)$ to $(Z, H(g_Z))$. Observing that for any $f \in k(y)^*$, with $y \in X^{(p-1)}$, the pair $(\operatorname{div} f, -[\log |f|^2])$ lies in $Z^p(\overline{X})$ by the Poincaré–Lelong formula, Theorem II.2, we obtain (i) since $(\operatorname{div} f, -[\log |f|^2])$ is mapped to $(\operatorname{div} f, -H[\log |f|^2])$.

Statement (ii) immediately follows from (i) and the definitions, while the proof of (iii) runs along the same lines as the proof of Theorem 1 (2).  □

**5.2**   In [A1], Arakelov introduced the group $CH^1(\overline{X})$, where $\overline{X} = (X, g_0)$ is an arithmetic surface and the metric $g_0$ on the Riemann surface $X(\mathbb{C})$ has Kähler form

$$\frac{i}{2g} \sum_{j=1}^{g} \omega_j \wedge \overline{\omega}_j,$$

where $g$ is the genus of $X(\mathbb{C})$ ($g \geq 1$) and $\omega_1, \ldots, \omega_g$ denotes an orthonormal basis of the space of holomorphic 1-forms $\Gamma(X(\mathbb{C}), \Omega^1_{X(\mathbb{C})})$.

**5.3**   Let $X, Y$ be regular schemes, projective and flat over $\mathbb{Z}$, let $f : Y \to X$ be a morphism such that $f_{\mathbb{Q}} : Y_{\mathbb{Q}} \to X_{\mathbb{Q}}$ is smooth and assume that $\delta = \dim Y - \dim X = d - 1$. Furthermore, let $\overline{L}_1, \ldots, \overline{L}_d$ be hermitian line bundles on $Y$ (cf. §4.2). By Theorem 2, Theorem 3 and Proposition 1, we get a class

$$f_*(\hat{c}_1(\overline{L}_1) \cdot \ldots \cdot \hat{c}_1(\overline{L}_d)) \in \widehat{CH}^1(X).$$

On the other hand, Deligne [D] and Elkik [El] constructed an hermitian line bundle $\langle \overline{L}_1, \ldots, \overline{L}_d \rangle$ on $X$. It seems likely that

$$\hat{c}_1(\langle \overline{L}_1, \cdots, \overline{L}_d \rangle) = f_*(\hat{c}_1(\overline{L}_1) \cdot \ldots \cdot \hat{c}_1(\overline{L}_d));$$

see [GS3] for the case $d = 2$.

For example, when $X = \operatorname{Spec} \mathbb{Z}$ and $d = 2$, if $L_i$ has a nonzero global section $s_i$ (for $i = 1, 2$) one gets (in $\widehat{CH}^1(X) \cong \mathbb{R}$)

$$f_*(\hat{c}_1(\overline{L}_1) \cdot \hat{c}_1(\overline{L}_2)) = \hat{c}_1(\langle \overline{L}_1, \overline{L}_2 \rangle) =$$

$$\sum_{y \in Y^{(2)}} \log \sharp(\mathcal{O}_{Y,y}/\langle s_1, s_2 \rangle) - \sum_{\alpha} \log \|s_2(P_\alpha)\| - \int_{Y(\mathbb{C})} \log \|s_1\| c_1(L_2, \|\cdot\|).$$

Here $\langle s_1, s_2 \rangle \subset \mathcal{O}_{Y,y}$ is the ideal generated by $s_1$ and $s_2$ after any choice of a trivialisation $L_{i,y} \cong \mathcal{O}_{Y,y}$, $\operatorname{div}(s_1|_{Y(\mathbb{C})}) = \sum n_\alpha P_\alpha$ (we assume that $\operatorname{div}(s_1)$ and $\operatorname{div}(s_2)$ intersect properly), and $\sharp$ denotes the cardinality of a finite set. The above formula coincides with Arakelov's intersection pairing given in [A1] (see also [L]), except for the last summand, which does not occur there, because Arakelov considered only line bundles with "admissible metrics", so the integral vanishes; see also [S3].

## 6. Heights

One use of the arithmetic intersection theory that we have described in this chapter is to provide a notion of *height* for projective varieties. Let $\mathbb{P}^N$ be the $N$-dimensional projective space over $\mathbb{Z}$, equipped with its standard Kähler structure; see §IV.1.1 below. Given $X \subset \mathbb{P}^N$ a closed irreducible subset of dimension $d+1$ ($d \geq -1$) we denote by $g_X$ a Green current for $X$ such that $dd^c g_X + \delta_X$ is harmonic and the harmonic projection $H(g_X)$ vanishes. From Theorem II.1, it follows that $g_X$ is unique modulo $(\text{im}\partial + \text{im}\overline{\partial})$. We let $\hat{X} \in \widehat{CH}^{N-d}(\mathbb{P}^N)$ be the class of the arithmetic cycle $(X, g_X)$. On the other hand, let $\hat{c}_1(\overline{\mathcal{O}(1)}) \in \widehat{CH}^1(\mathbb{P}^N)$ be the first arithmetic Chern class of the standard hermitian line bundle on $\mathbb{P}^N$ (see §IV.1.1 with $n = 1$ and $m = N$), and $f : \mathbb{P}^N \to \text{Spec}\,\mathbb{Z}$ the projection map. Faltings [F3] defined the height of $X$ by the formula

$$(8) \qquad h(X) = f_*(\hat{X} \cdot \hat{c}_1(\overline{\mathcal{O}(1)})^{d+1}) \in \mathbb{R} = \widehat{CH}^1(\text{Spec}\,\mathbb{Z}).$$

This formula is analogous to the classical formula formula for the *degree* of the generic fiber $X_{\mathbb{Q}} \subset \mathbb{P}^N_{\mathbb{Q}}$, using Chow groups, namely

$$(9) \qquad \deg(X_{\mathbb{Q}}) = f_*([X_{\mathbb{Q}}] \cdot c_1(\mathcal{O}(1))^d) \in \mathbb{Z} = CH^0(\text{Spec}\,\mathbb{Q}).$$

When $X_{\mathbb{Q}}$ is a rational point $P$, $h(X)$ coincides with the usual logarithmic height of $P$, i.e. $\log(\sqrt{x_0^2 + \cdots + x_N^2})$, where $(x_0, \cdots, x_N) \in \mathbb{Z}^{N+1}$ are integral homogenous coordinates of $P$ with no common divisor. Previous definitions of $h(X)$, using Chow coordinates, are due to Nesterenko and Philippon; see [P] and [S4] for a precise comparison of them.

Faltings [F3] proved that $h(X) \geq 0$. The stronger inequality

$$h(X) \geq (\sum_{k=1}^{d} \sum_{j=1}^{k} \frac{1}{2j}) \deg(X_{\mathbb{Q}})$$

holds, [BoGS], with equality only when $X$ is one of the projective subspace obtained by requiring that $N - d$ among the standard coordinates are equal to zero.

Also, if $X_{\mathbb{Q}}$ and $Y_{\mathbb{Q}}$ meet properly, the height of the cycle $X \cap Y$ (defined by Fulton's method [Fu]) is bounded above

$$(10) \quad h(X \cap Y) \leq h(X)\deg(Y_{\mathbb{Q}}) + \deg(X_{\mathbb{Q}})h(Y) + c\deg(X_{\mathbb{Q}})\deg(Y_{\mathbb{Q}}),$$

where $c$ is an explicit constant depending only on the dimensions involved [BoGS]. This inequality (10) constitutes some arithmetic analog of the Bézout theorem $\deg(X_{\mathbb{Q}} \cap Y_{\mathbb{Q}}) = \deg(X_{\mathbb{Q}}) \cdot \deg(Y_{\mathbb{Q}})$.

Faltings used $h(X)$ in his study of rational points on abelian varieties [F3].

# IV

## Characteristic Classes

We shall now define, following [GS3], characteristic classes with values in the arithmetic Chow groups that we introduced in the last Chapter.

If $E$ is a vector bundle on some arithmetic variety $X$, and $h$ an hermitian metric (invariant by complex conjugation) on the holomorphic bundle induced by $E$ on the set of complex points of $X$, there is a *Chern character* $\widehat{\mathrm{ch}}(E, h)$ in $\oplus_{p \geq 0} \widehat{CH}^p(X)_{\mathbb{Q}}$. It satisfies the usual axioms of a Chern character. For instance, it is multiplicative under tensor product and additive under orthogonal direct sums. But it does depend on the choice of the metric $h$. When $h$ is replaced by $h'$, the difference $\widehat{\mathrm{ch}}(E, h') - \widehat{\mathrm{ch}}(E, h)$ is the image in the arithmetic Chow groups of a secondary characteristic class introduced by Bott and Chern [BC], which played a role in Donaldson's work on Hermitian–Einstein metrics [Do]. More generally, we give in Proposition 1 a formula for the behavior of $\widehat{\mathrm{ch}}$ with respect to exact sequences.

In §1 we prove a splitting principle for the arithmetic Chow groups of Grassmannians, or rather their subgroups à la Arakelov, as in §III.5. This provides a definition of $\widehat{\mathrm{ch}}$ for the tautological bundles on these Grassmannians. In §2 we recall how any metric on a holomorphic vector bundle gives rise to form representatives of its characteristic classes. In §3 we define the Bott–Chern secondary characteristic classes by a method introduced in [BGS1] and [GS3]. In §4 we define $\widehat{\mathrm{ch}}$. For this we use the fact that any bundle on $X$, once tensored by an ample line bundle, is classified by a map to a Grassmannian. We then have to check that the class we get is independent of the choices. Another construction, due to Elkik [El], is sketched in §4.7. Finally, we give, following [GS3], a definition of the group of virtual hermitian vector bundles on $X$.

## 1. Arakelov Chow groups of the Grassmannians

**1.1**     Put $S := \operatorname{Spec}\mathbb{Z}$. For positive integers $m, n$, let

$$G_{m,n} := \operatorname{Grass}_n(\mathcal{O}_S^{m+n})$$

be the Grassmannian over $S$ representing the contravariant functor from the category of $S$-schemes to the category of sets which assigns to each $S$-scheme $T$ the set of locally free quotients of $\mathcal{O}_T^{m+n}$ of rank $n$ (cf. [DG], Théorème I.9.7.4). Assume $m = qn$ for $q$ a positive integer, and put $P := (G_{q,1})^n$; note that $G_{q,1} = \mathbb{P}_S(\mathcal{O}_S^{q+1})$ is the projective space bundle of $\mathcal{O}_S^{q+1}$ over $S$. There is a natural map $\mu : P \to G := G_{qn,n}$ coming from the direct sum of line bundles

$$(\mathcal{O}_T^{q+1} \longrightarrow \mathcal{L}_i : i = 1, \ldots, n) \longmapsto (\mathcal{O}_T^{qn+n} \longrightarrow \bigoplus_i \mathcal{L}_i).$$

We also note that the symmetric group $\Sigma_n$ acts on $P$ by permuting the factors.

On $G(\mathbb{C}) \cong U(m+n)/(U(m) \times U(n))$, where $U(m)$ denotes the unitary group of rank $m$, consider the Kähler form $\omega_G$ given locally by

$$\omega_G = -dd^c \log \|s\|^2,$$

where $s$ is a (local) non-vanishing section of the highest exterior power of the tautological quotient bundle of rank $n$ on $G(\mathbb{C})$. We note that $\omega_G$ is a Kähler form, that $F_\infty^* \omega_G = -\omega_G$, and that $\omega_G$ is invariant under the action of $U(m+n)$. The pair $\overline{G} = (G, \omega_G)$ defines an Arakelov variety, in the sense of Definition III.2. Analogously, we obtain the Arakelov variety $\overline{P} = (P, \omega_P)$, on which $\Sigma_n$ acts. We then have the following "splitting principle":

**Theorem 1**     *Assume $p \le q$. Then the natural map $\mu : P \to G$ induces an isomorphism*

(1)                    $$\mu^* : CH^p(\overline{G})_{\mathbb{Q}} \cong CH^p(\overline{P})_{\mathbb{Q}}^{\Sigma_n}$$

*of Arakelov Chow groups tensored with $\mathbb{Q}$, where $CH^p(\overline{P})^{\Sigma_n}$ denotes the invariants of $CH^p(\overline{P})$ under the action of $\Sigma_n$.*

**1.2**     To prove Theorem 1 we first note that the pull-back map

$$\mu^* : \widehat{CH}^p(G) \longrightarrow \widehat{CH}^p(P)$$

(see Theorem III.3) induces a map

$$\mu^* : CH^p(\overline{G}) \longrightarrow CH^p(\overline{P}).$$

Indeed harmonic forms on $G(\mathbb{C})$ are characterized by their invariance under the action of $U(m+n)$; this is a standard fact, which goes back to E.Cartan, see for instance [DR2] p.648. Whence their pull-backs to $P(\mathbb{C})$ are invariant under the action of $U(q+1)^n \subset U(m+n)$, and therefore they are also harmonic.

The exact sequence from Proposition III.2, applied to $G$ and $P$, gives us the following commutative diagram

$$
\begin{array}{ccccccccc}
CH^{p-1,p}(G)_{\mathbb{Q}} & \xrightarrow{\rho} & H^{p-1,p-1}(G)_{\mathbb{Q}} & \xrightarrow{a} & CH^p(\overline{G})_{\mathbb{Q}} & \xrightarrow{\varsigma} & CH^p(G)_{\mathbb{Q}} & \longrightarrow & 0 \\
\downarrow{\mu^*} & & \downarrow{\mu^*} & & \downarrow{\mu^*} & & \downarrow{\mu^*} & & \\
CH^{p-1,p}(P)_{\mathbb{Q}} & \xrightarrow{\rho} & H^{p-1,p-1}(P)_{\mathbb{Q}} & \xrightarrow{a} & CH^p(\overline{P})_{\mathbb{Q}} & \xrightarrow{\varsigma} & CH^p(P)_{\mathbb{Q}} & \longrightarrow & 0.
\end{array}
$$

We will now show that $CH^{p-1,p}(G)_{\mathbb{Q}} = CH^{p-1,p}(P)_{\mathbb{Q}} = 0$ and that $\mu^*$ induces isomorphisms $CH^p(G)_{\mathbb{Q}} \cong CH^p(P)_{\mathbb{Q}}^{\Sigma_n}$ and

$$
H^{p-1,p-1}(G)_{\mathbb{Q}} \cong H^{p-1,p-1}(P)_{\mathbb{Q}}^{\Sigma_n}.
$$

The theorem follows by the 5-lemma.

**1.3**    By Theorem I.3, $CH^p(G)_{\mathbb{Q}}$ (resp. $CH^{p-1,p}(G)_{\mathbb{Q}}$) is isomorphic to $K_0(G)^{(p)}$ (resp. $K_1(G)^{(p)}$), the weight $p$ part of $K_0(G)_{\mathbb{Q}}$ (resp. $K_1(G)_{\mathbb{Q}}$), where, for any $k > 0$, the Adams operation $\psi^k$ acts by multiplication by $k^p$. Similarly, $CH^p(P)_{\mathbb{Q}}$ is isomorphic to $K_0(P)^{(p)}$. Therefore we are reduced to a problem in $K$-theory. We now determine the structure of the $K$-groups $K_r(G)$ and $K_r(P)$.

On $E_0 := G$, we consider the tautological exact sequence

$$
0 \longrightarrow \mathcal{E} \longrightarrow \mathcal{O}_G^{m+n} \longrightarrow \mathcal{E}_0 \longrightarrow 0,
$$

where $\mathcal{E}_0$ is the tautological quotient sheaf, locally free of rank $n$. Let $E_1 := \mathbb{P}_{E_0}(\mathcal{E}_0)$ be the projective space bundle of the bundle $\mathcal{E}_0$ over $E_0$. On $E_1$ there is a tautological exact sequence

$$
0 \longrightarrow \mathcal{E}_1 \longrightarrow \pi_0^* \mathcal{E}_0 \longrightarrow \mathcal{L}_1 \longrightarrow 0,
$$

where $\mathcal{E}_1$ (resp. $\mathcal{L}_1$) is locally free of rank $n-1$ (resp. 1) and $\pi_0$ denotes the projection $E_1 \longrightarrow E_0$. Iterating this procedure, we get $E_{i+1} = \mathbb{P}_{E_i}(\mathcal{E}_i)$ with a tautological exact sequence

$$
0 \longrightarrow \mathcal{E}_{i+1} \longrightarrow \pi_i^* \mathcal{E}_i \longrightarrow \mathcal{L}_{i+1} \longrightarrow 0,
$$

where $\mathcal{E}_{i+1}$ (resp. $\mathcal{L}_{i+1}$) is locally free of rank $n-i-1$ (resp. 1) and $\pi_i$ denotes the projection $E_{i+1} \to E_i$. This construction ends with the scheme $E := E_n$ and the invertible sheaf $\mathcal{L}_n \cong \pi_{n-1}^* \mathcal{E}_{n-1}$. According to [Q1], Proposition 4.3, $K_r(E)$ can be computed as follows. One has $K_r(E_1) = \bigoplus_{j=0}^{n-1} K_r(E_0) \cdot \zeta^j$, where $\zeta$ denotes the class of $\mathcal{L}_1$ in $K_0(E_1)$.

Inductively this leads to

(2) $$K_r(E) = \bigoplus_{0 \leq \alpha_j \leq n-j} K_r(G) \cdot \zeta_1^{\alpha_1} \cdot \ldots \cdot \zeta_n^{\alpha_n},$$

where $\zeta_j$ denotes the class of the pull-back of $\mathcal{L}_j$ in $K_0(E)$ ($j = 1, \ldots, n$).

But the scheme $E$ can be viewed in a different way as follows. We start with $F_0 := S$ and the sheaf $\mathcal{F}_0 := \mathcal{O}_S^{m+n}$. Put $F_1 := \mathbb{P}_{F_0}(\mathcal{F}_0)$ with tautological exact sequence

$$0 \longrightarrow \mathcal{F}_1 \longrightarrow \rho_0^* \mathcal{F}_0 \longrightarrow \mathcal{M}_1 \longrightarrow 0,$$

where $\mathcal{F}_1$ (resp. $\mathcal{M}_1$) is locally free of rank $m + n - 1$ (resp. 1) and $\rho_0$ denotes the projection $F_1 \to F_0$. Iterating this procedure $n$ times, we get the scheme $F := F_n = \mathbb{P}_{F_{n-1}}(\mathcal{F}_{n-1})$ and the invertible sheaf $\mathcal{M}_n$. Clearly $E = F$ and the pull-backs of $\mathcal{L}_j$ and $\mathcal{M}_j$ to $E$ are isomorphic ($j = 1, \ldots, n$). By applying [Q1], Proposition 4.3, again, we obtain

(3)
$$K_r(E) = K_r(F) = \bigoplus_{0 \leq \beta_j \leq m+n-j} K_r(S) \cdot \zeta_1^{\beta_1} \cdot \ldots \cdot \zeta_n^{\beta_n}$$
$$= \bigoplus_{0 \leq \beta_j \leq m+n-j} K_r(\mathbb{Z}) \cdot \zeta_1^{\beta_1} \cdot \ldots \cdot \zeta_n^{\beta_n}.$$

Comparing (2) with (3), we immediately deduce $K_1(G)_\mathbb{Q} = 0$, because $K_1(\mathbb{Z}) \cong \mathbb{Z}/2\mathbb{Z}$. This in turn implies $CH^{p-1,p}(G)_\mathbb{Q} = 0$.

Because $G_{q,1} = \mathbb{P}_S(\mathcal{O}_S^{q+1})$, we obtain

(4) $$K_r(P) = \bigoplus_{0 \leq \gamma_j \leq q} K_r(\mathbb{Z}) \cdot \eta_1^{\gamma_1} \cdot \ldots \cdot \eta_n^{\gamma_n},$$

where $\eta_j$ denotes the class of the pull-back via the $j$-th projection $P \to G_{q,1}$ of the tautological invertible sheaf in $K_0(P)$. From (4) we derive, as before, $CH^{p-1,p}(P)_\mathbb{Q} = 0$.

**1.4**   We are now able to prove that $\mu^*$ induces an isomorphism

$$CH^p(G)_\mathbb{Q} \cong CH^p(P)_\mathbb{Q}^{\Sigma_n},$$

by showing that $K_0(G)^{(p)} \cong (K_0(P)^{(p)})^{\Sigma_n}$ for $p \leq q$. Because $p \leq nq \leq m + n - j$ and $p \leq q$, we have, by (3) and (4),

$$K_0(E)^{(p)} = \mathbb{Q}[\zeta_1, \ldots, \zeta_n]^{(p)},$$

and

$$K_0(P)^{(p)} = \mathbb{Q}[\eta_1, \ldots, \eta_n]^{(p)}.$$

Furthermore, the above construction of $E$ shows that the natural map

$\mu : P \to G$ factors through $E$, i.e.

$$
\begin{array}{ccc}
 & E & \\
\nu \nearrow & & \downarrow \pi \\
P & \overset{\mu}{\longrightarrow} & G
\end{array}
$$

and the induced map $\nu^* : K_0(E) \to K_0(P)$ takes $\zeta_j$ to $\eta_j$ $(j = 1, \ldots, n)$, whence we obtain an isomorphism

(5) $\qquad\qquad K_0(P)^{(p)} \cong K_0(E)^{(p)} = \mathbb{Q}[\zeta_1, \ldots, \zeta_n]^{(p)}.$

Note that $K_0(G)_{\mathbb{Q}}$ contains $\mathbb{Q}[\zeta_1, \ldots, \zeta_n]^{\Sigma_n}$, because we have $[\pi^*\mathcal{E}_0] = \bigoplus_{j=1}^n \zeta_j$ in $K_0(E)$, and $K_0(G)$ is a $\lambda$-ring. So we have the following inclusions, with $|\alpha| := \sum_j \alpha_j$,

$$
\bigoplus_{0 \le \alpha_j \le n - j} (\mathbb{Q}[\zeta_1, \ldots, \zeta_n]^{\Sigma_n})^{(p - |\alpha|)} \zeta_1^{\alpha_1} \cdot \ldots \cdot \zeta_n^{\alpha_n}
$$

(6) $\qquad \subseteq \bigoplus_{0 \le \alpha_j \le n - j} K_0(G)^{(p - |\alpha|)} \zeta_1^{\alpha_1} \cdot \ldots \cdot \zeta_n^{\alpha_n} \subseteq K_0(E)^{(p)}$

$$
= \mathbb{Q}[\zeta_1, \ldots, \zeta_n]^{(p)}.
$$

But now $\{\zeta_1^{\alpha_1} \cdot \ldots \cdot \zeta_n^{\alpha_n} : 0 \le \alpha_j \le n - j\}$ is a basis for $\mathbb{Q}[\zeta_1, \ldots, \zeta_n]$ over $\mathbb{Q}[\zeta_1, \ldots, \zeta_n]^{\Sigma_n}$, therefore all the inclusions in (6) are equalities. This shows that

$$
K_0(G)^{(p)} = (\mathbb{Q}[\zeta_1, \ldots, \zeta_n]^{\Sigma_n})^{(p)} = (\mathbb{Q}[\zeta_1, \ldots, \zeta_n]^{(p)})^{\Sigma_n} \cong (K_0(P)^{(p)})^{\Sigma_n},
$$

by the isomorphism (4), as claimed.

The above argument can be repeated if we replace Chow theory by cohomology, so we also have an isomorphism

(7) $\qquad\qquad H^{p-1,p-1}(G)_{\mathbb{Q}} \cong H^{p-1,p-1}(P)_{\mathbb{Q}}^{\Sigma_n}.$

This finishes the proof of Theorem 1. $\qquad\qquad\qquad\qquad\qquad\qquad \Box$

## 2. Chern forms

**2.1**    Let $X$ be a complex manifold, $E$ a holomorphic vector bundle of rank $r$ on $X$. Denote by $A^n(X, E)$ the smooth sections of $\Lambda^n T^* X \otimes E$, where $T^*(X)$ denotes the cotangent bundle of $X$; note that $A^n(X, E)$ is an $A^0(X)$-module where $A^0(X)$ are the $C^\infty$-functions on $X$. A *connection* on $E$ is a $\mathbb{C}$-linear map

$$
\nabla : A^0(X, E) \longrightarrow A^1(X, E)
$$

satisfying

$$\nabla(f \cdot s) = df \otimes s + f \cdot \nabla s$$

for all $f \in A^0(X)$, $s \in A^0(X, E)$. The decomposition $A^1(X, E) = A^{1,0}(X, E) \oplus A^{0,1}(X, E)$ induces a decomposition $\nabla = \nabla^{1,0} + \nabla^{0,1}$. A connection $\nabla$ gives also a map

$$\nabla : A^1(X, E) \longrightarrow A^2(X, E)$$

by setting

$$\nabla(\omega \otimes s) := d\omega \otimes s - \omega \otimes \nabla s \qquad (\omega \in A^1(X), \ s \in A^0(X, E)).$$

From this we deduce that

$$\begin{aligned}
\nabla^2(f \cdot s) &= \nabla(df \otimes s + f \cdot \nabla s) \\
&= d^2 f \otimes s - df \otimes \nabla s + \nabla(f \cdot \nabla s) \\
&= -df \otimes \nabla s + df \otimes \nabla s + f \cdot \nabla(\nabla s) = f \cdot \nabla^2(s).
\end{aligned}$$

In other words,

$$\nabla^2 : A^0(X, E) \longrightarrow A^2(X, E)$$

is an $A^0(X)$-linear map, hence $\nabla^2 \in A^2(X, \mathrm{End}(E))$; it is called the *curvature* of $\nabla$.

**2.2**   Let $U \subset X$ be an open set and $E|_U \cong \mathbf{C}^r$ a trivialization of $E$ on $U$. Given a section $s \in A^0(X, E)$, its restriction to $U$ can be written $s|_U = (f_1, \ldots, f_r)$ with $f_j \in A^0(U)$ $(j = 1, \ldots, r)$. Put

$$(\bar{\partial}_E s)|_U := (\bar{\partial} f_1, \ldots, \bar{\partial} f_r).$$

If on $V \subset X$ there is another trivialization for $E$ and $U \cap V \neq \emptyset$, we have $s|_V = (g_1, \ldots, g_r)$ with $g_j \in A^0(V)$ $(j = 1, \ldots, r)$. On $U \cap V$ we get

$$g_j = \sum_k \varphi_{jk} f_k,$$

with holomorphic transition functions $\varphi_{jk}$. Therefore $\bar{\partial} g_j = \sum_k \varphi_{jk} \bar{\partial} f_k$. So $\bar{\partial}_E$ extends to a global map from $A^{0,0}(X, E)$ to $A^{0,1}(X, E)$. It is called the *Cauchy–Riemann operator* of $E$.

**2.3**   Assume now that $E$ is equipped with an hermitian metric $h$, i.e. an hermitian inner product $\langle \cdot, \cdot \rangle$ on each fiber $E_x$, depending smoothly on the point $x \in X$. Given two sections $s, t \in A^0(X, E)$, the hermitian metric on $E$ defines a $C^\infty$-function $\langle s, t \rangle$ on $X$ by

$$\langle s, t \rangle(x) := \langle s(x), t(x) \rangle.$$

We may extend $\langle \cdot, \cdot \rangle$ by linearity to get a pairing

$$\langle \cdot, \cdot \rangle : A^1(X, E) \otimes A^0(X, E) \to A^1(X).$$

In other words, if $s, t \in A^0(X, E)$ and $\omega \in A^1(X)$, we define

$$\langle s \otimes \omega, t \rangle := \langle s, t \rangle \cdot \omega.$$

Similarly we get

$$\langle \cdot, \cdot \rangle : A^0(X, E) \otimes A^1(X, E) \to A^1(X).$$

**Definition 1**  *A connection* $\nabla : A^0(X, E) \to A^1(X, E)$ *is called unitary if it satisfies the equation*

$$d\langle s, t \rangle = \langle \nabla s, t \rangle + \langle s, \nabla t \rangle.$$

**2.4**

**Lemma 1**  *There exists a unique connection*

$$\nabla : A^0(X, E) \longrightarrow A^1(X, E)$$

*with the properties*

(i)   $\nabla^{0,1} = \bar{\partial}_E$,
(ii)  $\nabla$ *is unitary.*

*Proof.* We first show uniqueness. By (ii), we have

$$d\langle s, t \rangle = \langle \nabla s, t \rangle + \langle s, \nabla t \rangle,$$

hence, by (i),

$$\bar{\partial}\langle s, t \rangle = \langle \bar{\partial}_E s, t \rangle + \langle s, \nabla^{1,0} t \rangle.$$

But the last equation determines $\nabla^{1,0}$ uniquely. Namely

$$\langle s, \nabla^{1,0} t \rangle = \bar{\partial}\langle s, t \rangle - \langle \bar{\partial}_E s, t \rangle.$$

We can take this equation as a definition for $\nabla^{1,0}$. We are left to check that $\nabla := \nabla^{1,0} + \bar{\partial}_E$ defines indeed a connection. But for all $s, t \in A^0(X, E)$ and $f, g \in A^0(X)$, we have

$$\nabla^{0,1}(f \cdot s) = \bar{\partial}_E(f \cdot s) = \bar{\partial} f \otimes s + f \cdot \nabla^{0,1} s,$$

and

$$\begin{aligned}
\langle s, \nabla^{1,0}(g \cdot t) \rangle &= \bar{\partial}\langle s, g \cdot t \rangle - \langle \bar{\partial}_E s, g \cdot t \rangle \\
&= (\bar{\partial}\bar{g})\langle s, t \rangle + \bar{g}\, \bar{\partial}\langle s, t \rangle - \bar{g}\langle \bar{\partial}_E s, t \rangle \\
&= \langle s, \partial g \otimes t \rangle + \langle s, g \cdot \nabla^{1,0} t \rangle,
\end{aligned}$$

i.e.

$$\nabla^{1,0}(g \cdot t) = \partial g \otimes t + g \cdot \nabla^{1,0} t,$$

which completes the proof.      □

The connection $\nabla$ from Lemma 1 is called the *hermitian holomorphic* connection of $(E, h)$. Notice that

$$\nabla^2 = \nabla^{1,1} \in A^{1,1}(X, \text{End}(E)),$$

because $\nabla^{0,2} = \overline{\partial}_E^2 = 0$, and hence $\nabla^{2,0}$ vanishes by unitarity.

**2.5**

**Definition 2**   *Let $X$ be a complex manifold, $\overline{E} = (E, h)$ a holomorphic vector bundle on $X$ with an hermitian metric, and $\nabla$ its hermitian holomorphic connection. The Chern character form $\text{ch}(\overline{E}) = \text{ch}(E, h)$ for the pair $(E, h)$ is defined by*

$$\text{ch}(E, h) := \text{tr}_E \, \exp\left(\frac{-1}{2\pi i}\nabla^2\right) \in \bigoplus_{p \geq 0} A^{p,p}(X).$$

Here $\text{tr}_E$ is the trace in $\text{End}(E)$. These Chern character forms are $d$ and $d^c$ closed and satisfy the following properties (cf. [GH]):

(i)   $f^*\text{ch}(\overline{E}) = \text{ch}(f^*\overline{E})$, for every holomorphic map $f : Y \to X$ of complex manifolds;

(ii)  $\text{ch}(\overline{E} \oplus \overline{F}) = \text{ch}(\overline{E}) + \text{ch}(\overline{F})$, for all hermitian vector bundles $\overline{E}, \overline{F}$;

(iii) $\text{ch}(\overline{E} \otimes \overline{F}) = \text{ch}(\overline{E}) \cdot \text{ch}(\overline{F})$, for all hermitian vector bundles $\overline{E}, \overline{F}$;

(iv)  $\text{ch}(\overline{L}) = \exp(c_1(\overline{L}))$, for every hermitian line bundle $\overline{L}$.

The form $\text{ch}(\overline{E})$ depends on the choice of the metric on $E$, but not its cohomology class, as will follow from Theorem 2 below.

## 3. Bott–Chern secondary characteristic classes

**3.1**   The Chern character forms described in the previous section are *not* additive on exact sequences, as is the case for their cohomology classes. Let

$$\mathcal{E} : 0 \longrightarrow E' \longrightarrow E \longrightarrow E'' \longrightarrow 0$$

be an exact sequence of vector bundles with hermitian metrics $h', h, h''$ respectively. Instead of $(\mathcal{E}, h', h, h'')$ we shall just write $\overline{\mathcal{E}}$. We have the following

**Theorem 2**   *There is a unique way to attach to every sequence $\overline{\mathcal{E}}$ a form $\widetilde{\text{ch}}(\overline{\mathcal{E}})$ in*

$$\widetilde{A}(X) := \bigoplus_{p \geq 0} A^{p,p}(X)/(\text{im}\,\partial + \text{im}\,\overline{\partial})$$

*satisfying the following properties:*

(i)   $dd^c \widetilde{ch}(\overline{\mathcal{E}}) = ch(\overline{E}') - ch(\overline{E}) + ch(\overline{E}'')$;

(ii)  $f^* \widetilde{ch}(\overline{\mathcal{E}}) = \widetilde{ch}(f^* \overline{\mathcal{E}})$, *for every holomorphic map* $f : Y \to X$ *of complex manifolds;*

(iii) $\widetilde{ch}(\overline{\mathcal{E}}) = 0$, *if* $\overline{\mathcal{E}}$ *is split, i.e.* $(E, h) = (E' \oplus E'', h' \oplus h'')$ *and the maps in* $\mathcal{E}$ *are the obvious ones .*

**Definition 3**  *The form* $\widetilde{ch}(\overline{\mathcal{E}})$ *is called the Bott–Chern character form of* $\overline{\mathcal{E}}$ (cf. [BC ], [Do], [BGS1], [GS3]).

*Proof.* We first prove the uniqueness of $\widetilde{ch}$. Let $\mathcal{O}(1)$ be the standard line bundle of degree one with its Fubini–Study metric on $\mathbb{P}^1_{\mathbb{C}}$ and let $\sigma$ be a section of $\mathcal{O}(1)$ vanishing only at $\infty$. On $X \times \mathbb{P}^1_{\mathbb{C}}$, consider the bundle $E'(1) := E' \otimes \mathcal{O}(1)$ with hermitian metric induced by the metric on $E'$ and $\mathcal{O}(1)$. If $E' \to E'(1)$ is the map given by $\mathrm{id}_{E'} \otimes \sigma$, let $\widetilde{E} := (E \oplus E'(1))/E'$. We get an exact sequence

$$\widetilde{\mathcal{E}} : 0 \longrightarrow E'(1) \longrightarrow \widetilde{E} \longrightarrow E'' \longrightarrow 0.$$

For $z \in \mathbb{P}^1_{\mathbb{C}}$, denote by $i_z : X \to X \times \mathbb{P}^1_{\mathbb{C}}$ the map given by $i_z(x) = (x, z)$. When $z \neq \infty$, since $\sigma(z) \neq 0$, we get $i_z^* \widetilde{E} \cong E$ . On the other hand $i_\infty^* \widetilde{E} \cong E' \oplus E''$, because $\sigma(\infty) = 0$ and $\mathcal{O}(1)_\infty \cong \mathbb{C}$, whence $i_\infty^* E'(1) \cong E'$. Using a partition of unity, we can choose a hermitian metric $\widetilde{h}$ on $\widetilde{E}$ such that the isomorphisms $i_0^* \widetilde{E} \cong E$ and $i_\infty^* \widetilde{E} \cong E' \oplus E''$ become isometries, i.e. $i_0^* \widetilde{h} = h$ and $i_\infty^* \widetilde{h} = h' \oplus h''$. Denoting by $d = d_x + d_z$ the total differential on $X \times \mathbb{P}^1_{\mathbb{C}}$, where $d_x$ is the differential on $X$ and $d_z$ the differential on $\mathbb{P}^1_{\mathbb{C}}$, we consider the integral

$$I := \int_{\mathbb{P}^1_{\mathbb{C}}} dd^c \widetilde{ch}(\widetilde{\mathcal{E}}) \cdot \log |z|^2.$$

We now compute $I$ in two ways. First we obtain, by property (i) applied to $\widetilde{\mathcal{E}}$,

$$I = \int_{\mathbb{P}^1_{\mathbb{C}}} (ch(\overline{E}'(1)) - ch(\widetilde{E}, \widetilde{h}) + ch(\overline{E}'')) \log |z|^2$$

$$= ch(\overline{E}') \int_{\mathbb{P}^1_{\mathbb{C}}} ch(\overline{\mathcal{O}}(1)) \log |z|^2 - \int_{\mathbb{P}^1_{\mathbb{C}}} ch(\widetilde{E}, \widetilde{h}) \log |z|^2$$

$$+ ch(\overline{E}'') \int_{\mathbb{P}^1_{\mathbb{C}}} \log |z|^2$$

$$= - \int_{\mathbb{P}^1_{\mathbb{C}}} ch(\widetilde{E}, \widetilde{h}) \log |z|^2.$$

(The first integral vanishes since it changes sign when $z$ is replaced by

$1/z$). Secondly, we get, by Stokes' Theorem and by properties (ii) and (iii),

$$I \equiv \int_{\mathbf{P}^1_{\mathbf{C}}} \widetilde{\mathrm{ch}}(\widetilde{\mathcal{E}}) d_z d_z^c \log |z|^2 = \int_{\mathbf{P}^1_{\mathbf{C}}} \widetilde{\mathrm{ch}}(\widetilde{\mathcal{E}})(\delta_0 - \delta_\infty)$$

$$= i_0^* \widetilde{\mathrm{ch}}(\widetilde{\mathcal{E}}) - i_\infty^* \widetilde{\mathrm{ch}}(\widetilde{\mathcal{E}}) = \widetilde{\mathrm{ch}}(i_0^* \widetilde{\mathcal{E}}) - \widetilde{\mathrm{ch}}(i_\infty^* \widetilde{\mathcal{E}}) = \widetilde{\mathrm{ch}}(\overline{\mathcal{E}})$$

(modulo $(\mathrm{im}\partial + \mathrm{im}\overline{\partial})$). Hence $\widetilde{\mathrm{ch}}(\overline{\mathcal{E}})$ is uniquely determined modulo $(\mathrm{im}\partial + \mathrm{im}\overline{\partial})$ by the formula

(8) $$\widetilde{\mathrm{ch}}(\overline{\mathcal{E}}) \equiv - \int_{\mathbf{P}^1_{\mathbf{C}}} \mathrm{ch}(\widetilde{E}, \widetilde{h}) \log |z|^2.$$

To prove the existence of $\widetilde{\mathrm{ch}}$, we take the formula (8) above as a definition. We have to check that this definition does not depend on the choice of the hermitian metric $\widetilde{h}$ and satisfies properties (i), (ii) and (iii). We first verify that property (i) is fulfilled. Since taking direct image of forms commutes with differentiation, we have, as $\mathrm{ch}(\widetilde{E}, \widetilde{h})$ is $d$ and $d^c$ closed,

$$d_x d_x^c \widetilde{\mathrm{ch}}(\overline{\mathcal{E}}) \equiv - \int_{\mathbf{P}^1_{\mathbf{C}}} dd^c (\mathrm{ch}(\widetilde{E}, \widetilde{h}) \log |z|^2)$$

$$= - \int_{\mathbf{P}^1_{\mathbf{C}}} \mathrm{ch}(\widetilde{E}, \widetilde{h}) d_z d_z^c \log |z|^2$$

$$= - \int_{\mathbf{P}^1_{\mathbf{C}}} \mathrm{ch}(\widetilde{E}, \widetilde{h})(\delta_0 - \delta_\infty)$$

$$= - i_0^* \mathrm{ch}(\widetilde{E}, \widetilde{h}) + i_\infty^* \mathrm{ch}(\widetilde{E}, \widetilde{h})$$

$$= - \mathrm{ch}(\overline{E}) + \mathrm{ch}(\overline{E}') + \mathrm{ch}(\overline{E}''),$$

by the properties of Chern character forms (cf. 2.5).

Now we check the independence of $\widetilde{\mathrm{ch}}(\overline{\mathcal{E}})$ on the choice of the hermitian metric $\widetilde{h}$. Let $\widetilde{h}'$ be another hermitian metric on $\widetilde{E}$ satisfying $i_0^* \widetilde{h}' = h$ and $i_\infty^* \widetilde{h}' = h' \oplus h''$. The above calculation applied to the exact sequence

$$0 \longrightarrow \widetilde{E} \longrightarrow \widetilde{E} \longrightarrow 0$$

on $X \times \mathbf{P}^1_{\mathbf{C}}$, where the first (resp. second) $\widetilde{E}$ has hermitian metric $\widetilde{h}$ (resp. $\widetilde{h}'$), gives the formula

$$-\mathrm{ch}(\widetilde{E}, \widetilde{h}) + \mathrm{ch}(\widetilde{E}, \widetilde{h}') = - \int_{\mathbf{P}^1_{\mathbf{C}}} dd^c (\mathrm{ch}(\widetilde{\widetilde{E}}, \widetilde{\widetilde{h}}) \log |w|^2),$$

where $d = d_x + d_z + d_w$, the bundle $\widetilde{\widetilde{E}}$ on $X \times \mathbf{P}^1_{\mathbf{C}} \times \mathbf{P}^1_{\mathbf{C}}$ is the pull-back of the bundle $\widetilde{E}$, and the hermitian metric $\widetilde{\widetilde{h}}$ satisfies

$$\widetilde{\widetilde{h}}|_{z=0} = h, \quad \widetilde{\widetilde{h}}|_{z=\infty} = h' \oplus h'', \quad \widetilde{\widetilde{h}}|_{w=0} = \widetilde{h}, \quad \widetilde{\widetilde{h}}|_{w=\infty} = \widetilde{h}'.$$

One way to get such an $\widetilde{\widetilde{h}}$ is to take, if $w_1, w_2$ are homogeneous coordinates for $w$,

$$\widetilde{\widetilde{h}} = \frac{|w_1|^2 \widetilde{h} + |w_2|^2 \widetilde{h}'}{|w_1|^2 + |w_2|^2}.$$

Hence we obtain (modulo $(\operatorname{im}\partial + \operatorname{im}\overline{\partial})$)

$$\int_{\mathbf{P}^1_{\mathbf{C}}} (\operatorname{ch}(\widetilde{E}, \widetilde{h}) - \operatorname{ch}(\widetilde{E}, \widetilde{h}')) \log |z|^2$$

$$= \int_{\mathbf{P}^1_{\mathbf{C}}} \int_{\mathbf{P}^1_{\mathbf{C}}} dd^c (\operatorname{ch}(\widetilde{\widetilde{E}}, \widetilde{\widetilde{h}}) \log |w|^2) \log |z|^2$$

$$\equiv \int_{\mathbf{P}^1_{\mathbf{C}}} \int_{\mathbf{P}^1_{\mathbf{C}}} \operatorname{ch}(\widetilde{\widetilde{E}}, \widetilde{\widetilde{h}}) \log |w|^2 d_z d_z^c \log |z|^2$$

$$= \left( \int_{\mathbf{P}^1_{\mathbf{C}}} \operatorname{ch}(\widetilde{\widetilde{E}}, \widetilde{\widetilde{h}}) \log |w|^2 \right)_{z=0} - \left( \int_{\mathbf{P}^1_{\mathbf{C}}} \operatorname{ch}(\widetilde{\widetilde{E}}, \widetilde{\widetilde{h}}) \log |w|^2 \right)_{z=\infty}$$

$$= 0 \,,$$

because $\widetilde{\widetilde{h}}|_{z=0} = h$ and $\widetilde{\widetilde{h}}|_{z=\infty} = h' \oplus h''$ are independent of $w$.

Properties (ii) and (iii) now follow easily.     $\square$

**3.2**    Here are some properties of Bott–Chern character forms:

**3.2.1**    $\widetilde{\operatorname{ch}}(\overline{E}_1 \oplus \overline{E}_2) \equiv \widetilde{\operatorname{ch}}(\overline{E}_1) + \widetilde{\operatorname{ch}}(\overline{E}_2).$

*Proof.* By the construction in the proof of Theorem 2, we have

$$\widetilde{\operatorname{ch}}(\overline{E}_1 \oplus \overline{E}_2) = - \int_{\mathbf{P}^1_{\mathbf{C}}} \operatorname{ch}(E_1 \widetilde{\oplus} E_2, h_1 \widetilde{\oplus} h_2) \log |z|^2$$

$$\equiv - \int_{\mathbf{P}^1_{\mathbf{C}}} \operatorname{ch}(\widetilde{E}_1 \oplus \widetilde{E}_2, \widetilde{h}_1 \oplus \widetilde{h}_2) \log |z|^2$$

$$= \widetilde{\operatorname{ch}}(\overline{E}_1) + \widetilde{\operatorname{ch}}(\overline{E}_2).$$

**3.2.2**    $\widetilde{\operatorname{ch}}(\overline{E} \otimes \overline{F}) \equiv \widetilde{\operatorname{ch}}(\overline{E}) \cdot \operatorname{ch}(\overline{F}) \,.$

*Proof.* Again by the construction in the proof of Theorem 2, we have, given any hermitian vector bundle $\overline{F} = (F, g)$,

$$\widetilde{\operatorname{ch}}(\overline{E} \otimes \overline{F}) = - \int_{\mathbf{P}^1_{\mathbf{C}}} \operatorname{ch}(\widetilde{E} \otimes F, \widetilde{h} \otimes g) \log |z|^2$$

$$\equiv - \int_{\mathbf{P}^1_{\mathbf{C}}} \operatorname{ch}(\widetilde{E}, \widetilde{h}) \operatorname{ch}(F, g) \log |z|^2$$

$$= \widetilde{\operatorname{ch}}(\overline{E}) \cdot \operatorname{ch}(\overline{F}).$$

**3.2.3**   Let

$$
\begin{array}{ccccc}
0 & & 0 & & 0 \\
\downarrow & & \downarrow & & \downarrow \\
0 \longrightarrow E_1' \longrightarrow & E_1 & \longrightarrow E_1'' \longrightarrow 0 \\
\downarrow & & \downarrow & & \downarrow \\
0 \longrightarrow E_2' \longrightarrow & E_2 & \longrightarrow E_2'' \longrightarrow 0 \\
\downarrow & & \downarrow & & \downarrow \\
0 \longrightarrow E_3' \longrightarrow & E_3 & \longrightarrow E_3'' \longrightarrow 0 \\
\downarrow & & \downarrow & & \downarrow \\
0 & & 0 & & 0
\end{array}
$$

be a diagram where all rows $\mathcal{E}_i$ and all columns $\mathcal{F}_j$ are exact $(i,j = 1,2,3)$. Then we have

$$
(9) \qquad \sum_{i=1}^{3}(-1)^i \widetilde{\mathrm{ch}}(\overline{\mathcal{E}}_i) \equiv \sum_{j=1}^{3}(-1)^j \widetilde{\mathrm{ch}}(\overline{\mathcal{F}}_j).
$$

For the proof of this fact see [GS3]. Note the following special case. Take $E_1' = E_2' = E$ with hermitian metric $h_1$, $E_1 = E$ with hermitian metric $h_2$ and $E_2 = E$ with hermitian metric $h_3$, and take all the other terms to be zero. Then the above formula reads

$$
(10) \qquad \widetilde{\mathrm{ch}}(h_1, h_2) + \widetilde{\mathrm{ch}}(h_2, h_3) \equiv \widetilde{\mathrm{ch}}(h_1, h_3).
$$

Here $\widetilde{\mathrm{ch}}(h_1, h_2) := \widetilde{\mathrm{ch}}(\overline{\mathcal{E}})$, for

$$
\mathcal{E} : 0 \longrightarrow E \longrightarrow E \longrightarrow 0,
$$

where the map is the identity, $E'' = 0$, $E' = E$ has metric $h_1$, and $E$ has metric $h_2$. In particular

$$
(11) \qquad dd^c \widetilde{\mathrm{ch}}(h_1, h_2) = \mathrm{ch}(E, h_1) - \mathrm{ch}(E, h_2).
$$

**3.3**   The construction of Bott–Chern secondary characteristic classes is also valid for other characteristic classes like Chern classes, Todd classes etc... To define them, one replaces the Chern character form $\mathrm{ch}(\widetilde{E}, \widetilde{h})$ in formula (8) above by the forms representing these classes; see [GS3] for more details. For instance, the component of degree zero of $\widetilde{\mathrm{ch}}(h_1, h_2)$ is equal to $\tilde{c}_1(h_1, h_2)$. It coincides with the smooth function $\log(h_2/h_1)$, as can be seen by checking the axioms of Theorem 2 in degree zero; (i) follows from the Poincaré–Lelong formula.

## 4. Arithmetic Chern characters

**4.1**    Let $X$ be a regular scheme, projective and flat over $\mathbf{Z}$.
**Definition 4**    *An hermitian vector bundle* $\overline{E} = (E, h)$ *on* $X$ *is an algebraic vector bundle* $E$ *on* $X$ *such that the induced holomorphic vector bundle on* $X(\mathbf{C})$ *has an hermitian metric* $h$, *which is invariant under complex conjugation, i.e.* $F_\infty^*(h) = h$.

**Theorem 3**    *Let* $X$ *be a regular scheme, projective and flat over* $\mathbf{Z}$ *and* $\overline{E} = (E, h)$ *an hermitian vector bundle. Then there is a unique way to define a characteristic class*

$$\widehat{\mathrm{ch}}(\overline{E}) \in \widehat{CH}(X)_{\mathbf{Q}} = \bigoplus_{p \geq 0} \widehat{CH}^p(X)_{\mathbf{Q}}$$

*satisfying the following properties:*

(i)    $f^* \widehat{\mathrm{ch}}(\overline{E}) = \widehat{\mathrm{ch}}(f^* \overline{E})$, *for every morphism* $f : Y \to X$ *of regular schemes, projective and flat over* $\mathbf{Z}$;

(ii)    $\widehat{\mathrm{ch}}(\overline{E} \oplus \overline{F}) = \widehat{\mathrm{ch}}(\overline{E}) + \widehat{\mathrm{ch}}(\overline{F})$, *for all hermitian vector bundles* $\overline{E}, \overline{F}$;

(iii)    $\widehat{\mathrm{ch}}(\overline{E} \otimes \overline{F}) = \widehat{\mathrm{ch}}(\overline{E}) \cdot \widehat{\mathrm{ch}}(\overline{F})$, *for all hermitian vector bundles* $\overline{E}, \overline{F}$;

(iv)    $\widehat{\mathrm{ch}}(\overline{L}) = \exp(\hat{c}_1(\overline{L}))$, *for every hermitian line bundle* $\overline{L}$ *(with* $\hat{c}_1(\overline{L})$ *defined as in §III.4.2);*

(v)    $\omega(\widehat{\mathrm{ch}}(\overline{E})) = \mathrm{ch}(\overline{E})$ *is the classical Chern character form (see §2 ; the map* $\omega$ *was defined in Theorem III.1).*

We call $\widehat{\mathrm{ch}}(\overline{E})$ the arithmetic Chern character of the hermitian vector bundle $\overline{E}$.

**4.2**    We divide the proof of Theorem 3 into two parts. In the first part (4.2, 4.3) we show the uniqueness, while in the second part (4.4, 4.5) we prove the existence of the arithmetic Chern characters.

We first have to recall the following notation and results from §1. We defined $G = G_{m,n} = \mathrm{Grass}_n(\mathcal{O}_S^{m+n})$  $(S = \mathrm{Spec}\,\mathbf{Z}, \ m = qn)$, $P = (G_{q,1})^n$ and a map $\mu : P \to G$. Furthermore, $\overline{Q}_{m,n}$ is the tautological quotient bundle of rank $n$ on $G$ with quotient metric induced by the standard Euclidean metric $\sum_{j=1}^{m+n} |z_j|^2$ on $\mathbf{C}^{m+n}$. For $p \leq q$, $\mu$ induces an isomorphism (cf. Theorem 1)

(12)    $$\mu^* : CH^p(\overline{G})_{\mathbf{Q}} \cong CH^p(\overline{P})_{\mathbf{Q}}^{\Sigma_n}$$

and, by the proof of Theorem 1, we also know that

$$\mu^*(\overline{Q}_{m,n}) = \overline{L}_1 \oplus \ldots \oplus \overline{L}_n,$$

an orthogonal sum of hermitian line bundles $\overline{L}_j$ on $\overline{P}$   $(j = 1, \ldots, n)$.

We now turn to the proof of uniqueness. We will give a formula for $\widehat{\mathrm{ch}}(\overline{E})$. First, from the properties (i), (ii) and (iv) we get

$$\mu^* \widehat{\mathrm{ch}}(\overline{Q}_{m,n}) = \widehat{\mathrm{ch}}(\mu^* \overline{Q}_{m,n}) = \widehat{\mathrm{ch}}(\bigoplus_{j=1}^{n} \overline{L}_j)$$

$$= \sum_{j=1}^{n} \widehat{\mathrm{ch}}(\overline{L}_j) = \sum_{j=1}^{n} \exp(\hat{c}_1(\overline{L}_j)),$$

hence the $p$-th component is given by

$$\mu^* \widehat{\mathrm{ch}}_p(\overline{Q}_{m,n}) = \sum_{j=1}^{n} \frac{\hat{c}_1(\overline{L}_j)^p}{p!} \in CH^p(\overline{P})_{\mathbb{Q}}.$$

The latter sum is invariant under the action of $\Sigma_n$ on $CH^p(\overline{P})_{\mathbb{Q}}$. By the isomorphism (12), this leads to the following formula for $\widehat{\mathrm{ch}}_p(\overline{Q}_{m,n})$ when $q \geq p$:

(13) $\quad \widehat{\mathrm{ch}}_p(\overline{Q}_{m,n}) = \mu^{*-1}\left( \displaystyle\sum_{j=1}^{n} \frac{\hat{c}_1(\overline{L}_j)^p}{p!} \right) \in CH^p(\overline{G})_{\mathbb{Q}} \subset \widehat{CH}^p(G)_{\mathbb{Q}}.$

When $q + 1 \leq p$, the functoriality (i) allows one to get $\widehat{\mathrm{ch}}_p(\overline{Q}_{m,n})$ by restriction from a bigger Grassmannian (for which $q \geq p$). We then have

(14) $\quad\quad\quad\quad \widehat{\mathrm{ch}}(\overline{Q}_{m,n}) = \displaystyle\sum_{p \geq 0} \widehat{\mathrm{ch}}_p(\overline{Q}_{m,n}) \in \widehat{CH}(G)_{\mathbb{Q}}.$

Because $X$ is projective, there exists a line bundle $L$ on $X$ such that $E \otimes L^{-1}$ is generated by global sections, i.e. there is a surjection $\mathcal{O}_X^N \longrightarrow$ $\longrightarrow E \otimes L^{-1}$ for some integer $N$. We may assume that $N = m + n$ with $n = rk\, E$ and $m = qn$ $(q \geq 1)$. Because $G$ represents the functor which assigns to each $S$-scheme $T$ the set of locally free quotients of $\mathcal{O}_T^{m+n}$ of rank $n$, there is a morphism $f \colon X \to G$ corresponding to the surjection $\mathcal{O}_X^{m+n} \longrightarrow E \otimes L^{-1}$, i.e. $f^*(Q_{m,n}) \cong E \otimes L^{-1}$. Putting any metric on $L$, the isomorphism

$$E \cong f^*(Q_{m,n}) \otimes L$$

needs not be an isometry. Therefore we have to investigate how the difference

$$\widehat{\mathrm{ch}}(\overline{E}) - f^* \widehat{\mathrm{ch}}(\overline{Q}_{m,n}) \cdot \widehat{\mathrm{ch}}(\overline{L})$$

depends on the metric chosen on $L$.

**4.3**

**Lemma 2** *Let $x \in \widehat{CH}^p(X \times \mathbb{P}^1_{\mathbb{Z}})$ and denote by $i_z \colon X \to X \times \mathbb{P}^1_{\mathbb{Z}}$ the* *morphism given by* $i_z(x) = (x, z)$. *Then*

$$i_0^*(x) - i_\infty^*(x) = a\left(\int_{\mathbf{P}_{\mathbf{C}}^1} \omega(x) \cdot \log |z|^2\right)$$

*(for the definitions of the maps* $\omega : \widehat{CH}^p(X \times \mathbf{P}_{\mathbf{Z}}^1) \to A^{p,p}(X \times \mathbf{P}_{\mathbf{Z}}^1)$ *and* $a : \widetilde{A}^{p-1,p-1}(X) \to \widehat{CH}^p(X)$, *see Theorem III.1).*

*Proof.* Let $x$ be the class of $(Z, g_Z)$. We may assume that $Z$ is irreducible and flat over $\mathbf{Z}$, since the statement is well known for the Chow groups with supports in finite fibers. By the Moving Lemma over $\mathbf{Q}$, we can then assume that $Z$ is not contained in a component of div $z = (0) - (\infty)$, and therefore that it meets this divisor properly on $X \times \mathbf{P}_{\mathbf{Z}}^1$.

Denoting by $\pi : X \times \mathbf{P}_{\mathbf{Z}}^1 \to X$ the projection, we have

$$\begin{aligned}
0 &= \pi_*([(Z, g_Z)] \cdot [(\operatorname{div} z, -[\log |z|^2])]) \\
&= \pi_*([(Z_0 - Z_\infty, g_Z \wedge (\delta_0 - \delta_\infty) - [\omega(x) \log |z|^2])]) \\
&= \pi_*([(Z_0 - Z_\infty, g_{Z_0} - g_{Z_\infty} - [\omega(x) \log |z|^2])]) \\
&= i_0^*(x) - i_\infty^*(x) - a\left(\int_{\mathbf{P}_{\mathbf{C}}^1} \omega(x) \cdot \log |z|^2\right).
\end{aligned}$$

$\square$

**Proposition 1**    *Let*

$$\mathcal{E} : 0 \to E' \to E \to E'' \to 0$$

*be an exact sequence of vector bundles with hermitian metrics* $h', h, h''$ *and* $\overline{\mathcal{E}} = (\mathcal{E}, h', h, h'')$. *Then*

(15)          $\widehat{\operatorname{ch}}(\overline{E}') - \widehat{\operatorname{ch}}(\overline{E}) + \widehat{\operatorname{ch}}(\overline{E}'') = a(\widetilde{\operatorname{ch}}(\overline{\mathcal{E}}))$,

*where* $\widetilde{\operatorname{ch}}(\overline{\mathcal{E}})$ *is the Bott–Chern character form of* $\overline{\mathcal{E}}$.

*Proof.* As in the proof of Theorem 2 (working now over $\mathbf{Z}$ instead of $\mathbf{C}$), one constructs a bundle $\widetilde{E}$ with hermitian metric $\widetilde{h}$ on $X \times \mathbf{P}_{\mathbf{Z}}^1$ such that $i_0^* \widetilde{E} \cong E$, $i_\infty^* \widetilde{E} \cong E' \oplus E''$ and $i_0^* \widetilde{h} = h$, $i_\infty^* \widetilde{h} = h' \oplus h''$. Taking $x = \widehat{\operatorname{ch}}(\widetilde{E}, \widetilde{h})$ in Lemma 2, we get

$$i_0^* \widehat{\operatorname{ch}}(\widetilde{E}, \widetilde{h}) - i_\infty^* \widehat{\operatorname{ch}}(\widetilde{E}, \widetilde{h}) = a\left(\int_{\mathbf{P}_{\mathbf{C}}^1} \omega(\widehat{\operatorname{ch}}(\widetilde{E}, \widetilde{h})) \log |z|^2\right),$$

hence, by properties (i), (iii) and (v) and the definition (8) of the Bott–Chern character form,

$$\widehat{\operatorname{ch}}(\overline{E}) - \widehat{\operatorname{ch}}(\overline{E}') - \widehat{\operatorname{ch}}(\overline{E}'') = -a(\widetilde{\operatorname{ch}}(\overline{\mathcal{E}})).$$

$\square$

Applying Proposition 1 to the exact sequence

$$\mathcal{E} : 0 \to E \xrightarrow{\cong} f^*(Q_{m,n}) \otimes L \longrightarrow 0$$

with the hermitian metrics chosen as above, we get

(16)               $\widehat{\mathrm{ch}}(\overline{E}) = f^* \widehat{\mathrm{ch}}(\overline{Q}_{m,n}) \cdot \widehat{\mathrm{ch}}(\overline{L}) + a(\tilde{\mathrm{ch}}(\overline{\mathcal{E}}))$.

This proves the uniqueness of $\widehat{\mathrm{ch}}$.


**4.4**   We now take formula (16) as a definition of $\widehat{\mathrm{ch}}(\overline{E})$. We have to verify that this definition does not depend on the morphism $f$ and the line bundle $L$, and that it satisfies properties (i)–(v).

We first check that it is independent of $f$. For this we assume that $L = L' = \mathcal{O}_X$ and that we have two morphisms $f : X \longrightarrow G_{m,n}$ with $f^*(Q_{m,n}) \cong E$, corresponding to the surjection $\underline{f} : \mathcal{P} := \mathcal{O}_X^{m+n} \longrightarrow \longrightarrow E$, and $f' : X \longrightarrow G_{m',n}$ with $f'^*(Q_{m',n}) \cong E$, corresponding to the surjection $\underline{f}' : \mathcal{P}' := \mathcal{O}_X^{m'+n} \longrightarrow E$.

Because $\mathcal{P}$ and $\mathcal{P}'$ are free, we can choose morphisms $\alpha : \mathcal{P} \longrightarrow \mathcal{P}'$ and $\alpha' : \mathcal{P}' \longrightarrow \mathcal{P}$ such that $\underline{f} = \underline{f}' \circ \alpha$ and $\underline{f}' = \underline{f} \circ \alpha'$ . Putting

$$g := \begin{pmatrix} 1 - \alpha' \circ \alpha & \alpha' \\ -\alpha & 1 \end{pmatrix} \in \mathrm{Aut}(\mathcal{O}_X^{m''+n}),$$

where $m'' = m + m' + n$, one easily checks that the composite projection maps

$$\varphi : \mathcal{P} \oplus \mathcal{P}' \longrightarrow \mathcal{P} \xrightarrow{\underline{f}} E$$

and

$$\varphi' : \mathcal{P} \oplus \mathcal{P}' \longrightarrow \mathcal{P}' \xrightarrow{\underline{f}'} E$$

satisfy the relation $\varphi \circ g = \varphi'$. Because $\mathrm{Aut}(\mathcal{O}_X^{m''+n}) = GL_{m''+n}(R)$ with $R = \Gamma(X, \mathcal{O}_X)$, which is of finite type over $\mathbf{Z}$, the maps $\varphi$ and $\varphi'$ induce the following commutative diagram

$$\begin{array}{ccc} X_R & \xrightarrow{\varphi} & G_R \\ & \varphi' \searrow & \downarrow g \\ & & G_R \, , \end{array}$$

with $X_R = X \otimes_{\mathbf{Z}} R$, $G_R = G_{m'',n} \otimes_{\mathbf{Z}} R$. The isomorphism $f^*(Q_{m,n}) \cong f'^*(Q_{m',n})$ now implies

$$\varphi^* g^*(Q_{m'',n}) \cong \varphi'^*(Q_{m'',n}).$$

**Lemma 3**    *In $\widehat{CH}(G_R)_{\mathbb{Q}}$ we have the identity*

$$\widehat{\mathrm{ch}}(\overline{Q}_{m'',n}) - g^*\widehat{\mathrm{ch}}(\overline{Q}_{m'',n}) = a(\widetilde{\mathrm{ch}}(\overline{Q}_{m'',n}, g^*\overline{Q}_{m'',n})),$$

*where the right-hand side is the Bott–Chern character form for the exact sequence $0 \to Q_{m'',n} \to g^*Q_{m'',n} \to 0$, with the standard metric on $Q_{m'',n}$ and the pull-back via $g$ of the standard metric on $g^*Q_{m'',n}$.*

*Proof.* For any $g \in GL_{m''+n}(R)$, we define

$$\delta(g) := \widehat{\mathrm{ch}}(\overline{Q}_{m'',n}) - g^*\widehat{\mathrm{ch}}(\overline{Q}_{m'',n}) - a(\widetilde{\mathrm{ch}}(\overline{Q}_{m'',n}, g^*\overline{Q}_{m'',n})) \in \widehat{CH}(G_R)_{\mathbb{Q}}.$$

We have

$$\zeta(\delta(g)) = \mathrm{ch}(Q_{m'',n}) - g^*\mathrm{ch}(Q_{m'',n}) = 0,$$

since the induced action of $GL_{m'',n}(R)$ on $CH(G_R)_{\mathbb{Q}}$ is trivial. Furthermore, we have

$$\omega(\delta(g)) = \mathrm{ch}(\overline{Q}_{m'',n}) - g^*\mathrm{ch}(\overline{Q}_{m'',n}) - dd^c\widetilde{\mathrm{ch}}(\overline{Q}_{m'',n}, g^*\overline{Q}_{m'',n}) = 0,$$

by Theorem 2. Therefore

$$\delta(g) \in \ker \zeta \cap \ker \omega = H^{p-1,p-1}(G_R)/\mathrm{im}\rho,$$

by Theorem III.1.

For any $g, h \in GL_{m''+n}(R)$ we get by direct computation, using Theorem 2 and 3.2.3,

$$\delta(g) + g^*\delta(h)$$
$$= \widehat{\mathrm{ch}}(\overline{Q}_{m'',n}) - g^*h^*\widehat{\mathrm{ch}}(\overline{Q}_{m'',n})$$
$$\quad - a(\widetilde{\mathrm{ch}}(\overline{Q}_{m'',n}, g^*\overline{Q}_{m'',n})) - a(\widetilde{\mathrm{ch}}(g^*\overline{Q}_{m'',n}, g^*h^*\overline{Q}_{m'',n}))$$
$$= \widehat{\mathrm{ch}}(\overline{Q}_{m'',n}) - (hg)^*\widehat{\mathrm{ch}}(\overline{Q}_{m'',n}) - a(\widetilde{\mathrm{ch}}(\overline{Q}_{m'',n}, (hg)^*\overline{Q}_{m'',n}))$$
$$= \delta(hg).$$

Because $GL_{m''+n}(\mathbb{C})$ is connected, $GL_{m''+n}(R)$ acts trivially on $H^{p-1,p-1}(G_R)$, whence we get

$$\delta(h) + \delta(g) = \delta(hg),$$

which implies in particular that the image via $\delta$ of a commutator vanishes in $\widehat{CH}(G_R)_{\mathbb{Q}}$.

Now we consider the natural map

$$j : G = G_{m'',n} \longrightarrow G_{2m'',2n} = G',$$

arising from the imbedding

$$j : GL_{m''+n}(R) \longrightarrow GL_{2m''+2n}(R),$$

$$g \longmapsto \begin{pmatrix} g & 0 \\ 0 & 1 \end{pmatrix}$$

and the induced homomorphism

$$j^* : H^{p-1,p-1}(G'_R) \longrightarrow H^{p-1,p-1}(G_R).$$

Taking $h = \begin{pmatrix} g & 0 \\ 0 & g^{-1} \end{pmatrix} \in GL_{2m''+2n}(R)$, one easily checks that $\delta(g) = j^*\delta(h)$. But $h$ is a commutator, because

$$h = \begin{pmatrix} g & 0 \\ 0 & 1 \end{pmatrix}\begin{pmatrix} 0 & 1 \\ 1 & 0 \end{pmatrix}\begin{pmatrix} g^{-1} & 0 \\ 0 & 1 \end{pmatrix}\begin{pmatrix} 0 & 1 \\ 1 & 0 \end{pmatrix},$$

hence $\delta(h) = 0$, as mentioned above. Therefore we find $\delta(g) = 0$, which completes the proof of Lemma 3. $\qquad\square$

Applying now $\varphi^*$ to the identity of Lemma 3, we derive

$$f^*\widehat{\mathrm{ch}}(\overline{Q}_{m,n}) - f'^*\widehat{\mathrm{ch}}(\overline{Q}_{m',n}) = a(\widetilde{\mathrm{ch}}(f^*\overline{Q}_{m,n}, f'^*\overline{Q}_{m',n})).$$

This implies, using 3.2.2 and 3.2.3, that $\widehat{\mathrm{ch}}(\overline{E})$, when defined by formula (16), is independent on the choice of $f$.

**4.5** Let us prove now that the definition of $\widehat{\mathrm{ch}}(\overline{E})$ by (16) does not depend on the choice of $L$. Assume we had $f : X \longrightarrow G_{m,n}$ with $f^*(Q_{m,n}) \cong E \otimes L^{-1}$, and $f' : X \longrightarrow G_{m',1}$ with $f'^*(Q_{m',1}) \cong L \otimes L'^{-1}$.

The composition of the product map $f \times f' : X \longrightarrow G_{m,n} \times G_{m',1}$ and the natural map $\nu : G_{m,n} \times G_{m',1} \longrightarrow G_{m'',n}$, given by taking tensor products (with $m'' = mm' + nm' + m$), then satisfies

$$(f \times f')^*\nu^*(Q_{m'',n}) = f^*(Q_{m,n}) \otimes f'^*(Q_{m',1}) \cong E \otimes L'^{-1}.$$

**Lemma 4**   Let $m'' = mm' + mn' + m'n$ and $n'' = nn'$. Let

$$\nu : G_{m,n} \times G_{m',n'} \longrightarrow G_{m'',n''},$$

be given by the tensor product. We have

$$\nu^*\widehat{\mathrm{ch}}(\overline{Q}_{m'',n''}) = \widehat{\mathrm{ch}}(\overline{Q}_{m,n}) \cdot \widehat{\mathrm{ch}}(\overline{Q}_{m',n'}).$$

*Proof.* By assumption, we have $m = qn, m' = qn'$, hence $m'' = qn''$. With the notations of §1, we obtain the following commutative diagram

$$
\begin{array}{ccc}
(G_{q,1})^n \times (G_{q,1})^{n'} & \xrightarrow{\text{id}} & (G_{q,1})^{n''} \\
{\scriptstyle \mu \times \mu'}\downarrow & & \downarrow{\scriptstyle \mu''} \\
G_{m,n} \times G_{m',n'} & \xrightarrow{\nu} & G_{m'',n''}.
\end{array}
$$

By the splitting principle, one checks that

$$(\mu \times \mu')^*(\nu^*\widehat{\mathrm{ch}}(\overline{Q}_{m'',n''}) - \widehat{\mathrm{ch}}(\overline{Q}_{m,n}) \cdot \widehat{\mathrm{ch}}(\overline{Q}_{m',n'})) = 0,$$

hence the lemma follows, because $(\mu \times \mu')^*$ is injective by Theorem 1.

$\qquad\square$

Now we obtain, using Lemma 4,

$$f^*\widehat{\mathrm{ch}}(\overline{Q}_{m,n})\widehat{\mathrm{ch}}(\overline{L}) + a(\widetilde{\mathrm{ch}}(\overline{E}, f^*\overline{Q}_{m,n} \otimes \overline{L}))$$

$$= f^*\widehat{\mathrm{ch}}(\overline{Q}_{m,n})\widehat{\mathrm{ch}}(f'^*\overline{Q}_{m',1} \otimes \overline{L}') + a(\widetilde{\mathrm{ch}}(\overline{E}, f^*\overline{Q}_{m,n} \otimes f'^*\overline{Q}_{m',1} \otimes \overline{L}'))$$

$$= (f \times f')^*\nu^*\widehat{\mathrm{ch}}(\overline{Q}_{m'',n})\widehat{\mathrm{ch}}(\overline{L}') + a(\widetilde{\mathrm{ch}}(\overline{E}, (f \times f')^*\nu^*\overline{Q}_{m'',n} \otimes \overline{L}')),$$

and this shows the independence of $\widehat{\mathrm{ch}}(\overline{E})$ on the choice of $L$.

Properties (i)–(v) follow from our definition of $\widehat{\mathrm{ch}}(\overline{E})$ in formula (16) and Lemma 4 .


**4.6**    More generally, given an hermitian vector bundle $\overline{E}$ of rank $r$ and $\varphi(T_1, \dots, T_r) \in \mathbb{Q}[[T_1, \dots, T_r]]$ a symmetric power series in $r$ variables, one may define a characteristic class

$$\widehat{\varphi}(\overline{E}) \in \widehat{CH}(X)_{\mathbb{Q}}$$

satisfying axioms similar to those of Theorem 3. In particular, when $\overline{E} = \bigoplus_{j=1}^{r} \overline{L}_j$ is an orthogonal direct sum of hermitian line bundles, it is given by the formula

$$\widehat{\varphi}(\overline{E}) = \varphi(\hat{c}_1(\overline{L}_1), \dots, \hat{c}_1(\overline{L}_r)).$$

For instance there are Chern classes $\hat{c}_p(\overline{E}) \in \widehat{CH}^p(X)_{\mathbb{Q}}$ — in fact $\hat{c}_p(\overline{E}) \in \widehat{CH}^p(X)$ — and a Todd class $\widehat{Td}(\overline{E}) \in \widehat{CH}(X)_{\mathbb{Q}}$.

To define $\widehat{\varphi}$ one may either mimick the construction of $\widehat{\mathrm{ch}}$ or consider the power series $\psi$ such that

$$\varphi(T_1, \dots, T_r) = \psi(U_1, \dots, U_p, \dots),$$

where

$$U_p = \bigoplus_{j=1}^{r} T_j^p / p!.$$

Then, if $\widehat{\mathrm{ch}}_p(\overline{E})$ is the component of degree $p$ of $\widehat{\mathrm{ch}}(\overline{E})$, one can take

$$\widehat{\varphi}(\overline{E}) = \psi(\widehat{\mathrm{ch}}_1(\overline{E}), \dots, \widehat{\mathrm{ch}}_p(\overline{E}), \dots).$$


**4.7**    Another approach to characteristic classes is an arithmetic analog of Segre classes due to Elkik [El]. Let $p : \mathbb{P}_X(E) \to X$ be the projective bundle attached to $E$. On $\mathbb{P}_X(E)$ we have a canonical exact sequence

$$\mathcal{E} : 0 \longrightarrow H \longrightarrow p^*(E) \longrightarrow L \longrightarrow 0,$$

where $H$ (resp. $L$) has rank $r - 1$ (resp. $1$). We endow $H$ and $L$ with the metrics induced from $p^*(\overline{E})$. Let

$$\hat{s}'_k(\overline{E}) = p_*(\hat{c}_1(\overline{L})^{r-1+k}) \in \widehat{CH}^k(X)$$

and define $R_k \in \widetilde{A}^{k-1,k-1}(X)$ by the following identity of power series in one variable $t$:

$$\sum_{k>0} R_k t^k = (\sum_{k>0} t^k p_*(c_1(\overline{L})^{k-1} \Delta))(\sum_{k \geq 0} c_k(\overline{E})(-t)^k)^{-1}.$$

Here $c_k(\overline{E}) \in A^{k,k}(X)$ are the Chern forms of $\overline{E}$, $c_1(\overline{L}) \in A^{1,1}(\mathbb{P}_X(E))$ is the first Chern form of $\overline{L}$, and $\Delta \in \widetilde{A}^{r-1,r-1}(\mathbb{P}_X(E))$ is the Bott–Chern class $\Delta = \tilde{c}_r(\mathcal{E}^* \otimes L)$ of the dual of $\mathcal{E}$ tensored with $L$. In particular

$$dd^c(\Delta) = -c_r(p^*(\overline{E})^* \otimes \overline{L}).$$

(Notice that $\widetilde{A}(X)$ is a module over closed forms in $A(X)$.)

The arithmetic Segre classes of $\overline{E}$ are $\hat{s}_0(\overline{E}) = 1$ and, when $k > 0$,

$$\hat{s}_k(\overline{E}) = \hat{s}'_k(\overline{E}) + a(R_k) \in \widehat{CH}^k(X).$$

The Chern classes $\hat{c}_p(\overline{E}) \in \widehat{CH}^p(X)$ can be defined by the identity of power series over the Chow ring

$$\sum_{k \geq 0} \hat{c}_k(\overline{E})(-t)^k = (\sum_{k \geq 0} \hat{s}_k(\overline{E})t^k)^{-1}.$$

The fact that this definition coincides with the one in 4.6 is a consequence of the properties of $\hat{c}_k$ (see [GS8]).

**4.8**

**Definition 5** *We denote by $\widehat{K}_0(X)$ the group generated by triples $(E, h, \eta)$, where $(E, h)$ is an hermitian vector bundle on $X$ and $\eta \in \bigoplus_{p \geq 0} \widetilde{A}^{p,p}(X)$, subject to the following relation*

$$(E', h', \eta') + (E'', h'', \eta'') = (E, h, \eta' + \eta'' + \widetilde{\mathrm{ch}}(\overline{\mathcal{E}})),$$

*for every exact sequence $\mathcal{E} : 0 \longrightarrow E' \longrightarrow E \longrightarrow E'' \longrightarrow 0$, and $\overline{\mathcal{E}} = (\mathcal{E}, h', h, h'')$ as in 3.1.*

**Theorem 4** *There is an isomorphism $\widehat{\mathrm{ch}} : \widehat{K}_0(X)_{\mathbb{Q}} \longrightarrow \widehat{CH}(X)_{\mathbb{Q}}$, given by*

$$\widehat{\mathrm{ch}}(E, h, \eta) := \widehat{\mathrm{ch}}(E, h) + a(\eta).$$

We shall not prove Theorem 4. We just mention that $\widehat{\mathrm{ch}}$ is well-defined by Proposition 1 and that Theorem 4 follows by the 5-lemma from the following diagram

$$
\begin{array}{ccccccccc}
K_1(X)_{\mathbb{Q}} & \longrightarrow & \bigoplus_{p \geq 0} \widetilde{A}^{p,p}(X) & \longrightarrow & \widehat{K}_0(X)_{\mathbb{Q}} & \longrightarrow & K_0(X)_{\mathbb{Q}} & \longrightarrow & 0 \\
\downarrow \cong & & \downarrow \mathrm{id} & & \downarrow \widehat{\mathrm{ch}} & & \downarrow \cong & & \\
\bigoplus_{p \geq 0} CH^{p-1,p}(X)_{\mathbb{Q}} & \longrightarrow & \bigoplus_{p \geq 0} \widetilde{A}^{p,p}(X) & \longrightarrow & \widehat{CH}(X)_{\mathbb{Q}} & \longrightarrow & CH(X)_{\mathbb{Q}} & \longrightarrow & 0,
\end{array}
$$

where the hard part is the proof of the commutativity of the left-hand square (see [GS3] and [Wa]).

# V

---

# The Determinant of Laplace Operators

Our aim is now to define a notion of *direct image* for hermitian vector bundles on arithmetic varieties. This will be done in the next chapter, and requires a notion of determinant for Laplace operators, which is described in this chapter.

Namely, given an hermitian vector bundle $E$ on a compact Kähler manifold, we consider the Laplace operator $\Delta$ acting on $E$-valued differential forms of a fixed degree $(0, q)$, $q \geq 0$. This operator has a discrete and non-negative spectrum. But, since its eigenvalues are unbounded, the definition of its determinant requires some *regularization* procedure. The method we shall be using is known as zeta function regularization: the determinant is defined as

$$\det{}'(\Delta) = \exp(-\zeta'_\Delta(0)),$$

where $\zeta_\Delta(s)$, $s \in \mathbb{C}$, is the zeta function of $\Delta$, defined by analytic continution from a half plane $\mathrm{Re}\, s > d$. This definition is due to Ray and Singer [RS].

In §1, we describe this regularization procedure in an axiomatic way, for infinite sequences of positive real numbers. Then we define generalized Laplacians in §2, and we sketch the construction of a heat kernel for these operators in §3; here we follow [BGV], and the reader is referred to this book for more details. We show in §4 that $\zeta_\Delta(s)$, the Mellin transform of the heat kernel of $\Delta$, has the required properties to make sense of $\det'(\Delta)$, the determinant of the restriction of $\Delta$ to the orthogonal complement to its kernel.

## 1. Regularization of infinite products

**1.1**    We first make sense of the following formula:

(1) $$1 \cdot 2 \cdot 3 \cdot 4 \cdots = \sqrt{2\pi} \quad .$$

**First Approach**    We look at Stirling's formula

$$n! = n^n \sqrt{n} e^{-n} \sqrt{2\pi} \left(1 + O(\frac{1}{n})\right),$$

i.e.

(2) $$\log(n!) = n \cdot \log n + \frac{1}{2} \log n - n + \log \sqrt{2\pi} + O(\frac{1}{n})$$

for $n \to \infty$ (cf. [WW], pp. 251-253).

**Definition 1**    *We define* $\log(\infty!)$ *to be the finite part of the expansion (2), hence*

$$\infty! := \sqrt{2\pi}.$$

**Second Approach**    Here we start with Riemann's zeta-function:

$$\zeta(s) := \sum_{n \geq 1} n^{-s} \quad (\operatorname{Re} s > 1).$$

Recall that $\zeta(s)$ has a meromorphic continuation to the whole complex plane, with a single pole at 1. More precisely:

(3) $$\zeta(s) = \frac{1}{s-1} + \gamma + O(s-1)$$

where $\gamma$ denotes Euler's constant (cf. [C], Chapter 2). Furthermore $\zeta(s)$ satisfies the following functional equation:

(4) $$\pi^{-s/2}\Gamma(s/2)\zeta(s) = \pi^{-\frac{(1-s)}{2}}\Gamma(\frac{1-s}{2})\zeta(1-s).$$

Since

$$\zeta'(s) = -\sum_{n \geq 1} \log(n) n^{-s}$$

when $\operatorname{Re} s > 1$, we are led to the following:

**Definition 2**    $\infty! := e^{-\zeta'(0)}.$

**Proposition 1**    $e^{-\zeta'(0)} = \sqrt{2\pi}$, *i.e. Definitions 1 and 2 are compatible.*

*Proof.* By the functional equation (4), we obtain

$$\zeta(s) = \pi^{s-1/2}\frac{\Gamma((1-s)/2)}{\Gamma(s/2)}\,\zeta(1-s),$$

hence

(5)   $$\frac{\zeta'(s)}{\zeta(s)} = \log\pi - \frac{1}{2}\cdot\frac{\Gamma'((1-s)/2)}{\Gamma((1-s)/2)} - \frac{1}{2}\cdot\frac{\Gamma'(s/2)}{\Gamma(s/2)} - \frac{\zeta'(1-s)}{\zeta(1-s)}.$$

We now compute the right-hand side of (5) for $s \to 0$. First we note from (3)

(6)   $$\frac{\zeta'(1-s)}{\zeta(1-s)} = \frac{1}{s} + \gamma + O(s) \qquad (s \to 0).$$

By the duplication formula for the $\Gamma$-function, we obtain (cf. [WW], p. 240)

$$\Gamma(s+1) = \pi^{-1/2}2^s\,\Gamma((s+1)/2)\Gamma(s/2+1),$$
$$\frac{\Gamma'(1)}{\Gamma(1)} = \log 2 + \frac{1}{2}\cdot\frac{\Gamma'(1/2)}{\Gamma(1/2)} + \frac{1}{2}\cdot\frac{\Gamma'(1)}{\Gamma(1)},$$

and

(7)   $$-\frac{1}{2}\cdot\frac{\Gamma'(1/2)}{\Gamma(1/2)} = \log 2 + \frac{\gamma}{2} \quad (\text{because } \gamma = -\Gamma'(1)).$$

Finally, we derive from $\Gamma(s) = s^{-1}\Gamma(s+1)$ that

(8)   $$\frac{\Gamma'(s)}{\Gamma(s)} = -\frac{1}{s} + \frac{\Gamma'(s+1)}{\Gamma(s+1)} = -\frac{1}{s} + \frac{\Gamma'(1)}{\Gamma(1)} + O(s) \qquad (s \to 0).$$

Adding (5), (6), (7), (8) and letting $s \to 0$, we obtain

$$\frac{\zeta'(0)}{\zeta(0)} = \log\pi + \log 2.$$

By the functional equation (4), we get $\zeta(0) = -1/2$, which leads to

$$\zeta'(0) = -\frac{1}{2}\,\log(2\pi),$$

hence

$$e^{-\zeta'(0)} = \sqrt{2\pi}.$$

$\square$

**Third Approach**   We start with the $\Theta$-function:

$$\Theta(t) := \sum_{n\geq 1} e^{-n^2 t}.$$

Taking the Mellin transform of $\Theta(t)$, we have, for $\mathrm{Re}\,s > 1$,

(9)   $$\zeta(2s) = \frac{1}{\Gamma(s)}\int_0^\infty \Theta(t)t^{s-1}dt.$$

If we could exchange the order of differentiation and integration on the right-hand side, we would get

$$2\zeta'(0) \text{ `` = ''} \int_0^\infty \Theta(t) \left( \frac{t^s}{\Gamma(s)} \right)'_{s=0} \frac{dt}{t} \text{ `` = ''} \int_0^\infty \Theta(t) \frac{dt}{t}.$$

Unfortunately, the latter integral does not exist, because the integrand diverges at $t = 0$. Indeed, it follows from the Poisson summation formula (cf. [C], p. 30), that

$$\Theta(t) = \frac{\sqrt{\pi}}{\sqrt{t}} \left( \Theta(\frac{\pi^2}{t}) + \frac{1}{2} \right) - \frac{1}{2},$$

hence

$$\Theta(t) = \frac{\sqrt{\pi}}{2\sqrt{t}} - \frac{1}{2} + O(\sqrt{t}) \qquad (t \to 0).$$

This implies, as shown in Theorem 1 below, that there exist some constants $a_0$ and $a_{-1}$, such that, as $\epsilon$ tends to zero, the function

$$\int_\epsilon^\infty \Theta(t) \frac{dt}{t} + a_0 \log \epsilon - a_{-1}/\sqrt{\epsilon}$$

has a limit, called the *finite part* of the integral. Notice that it is well defined since $a_0$ and $a_{-1}$ are unique. This suggests the following

**Definition 3**     *We define* $\log(\infty!)$ *to be the finite part of the integral*

$$-\frac{1}{2} \int_\epsilon^\infty \Theta(t) \frac{dt}{t}$$

*as $\epsilon$ tends to zero.*

**Proposition 2**     *We have*

$$2\zeta'(0) = \text{finite part of } \left( \int_\epsilon^\infty \Theta(t) \frac{dt}{t} \right)_{\epsilon \to 0} - \frac{\gamma}{2},$$

*i.e. Definitions 2 and 3 are NOT compatible.*

*Proof.* This is a special case of Theorem 1 below (with $\epsilon^2$ instead of $\epsilon$). $\qquad \qquad \square$

**1.2**     We shall now extend the second and third approaches in §1.1 to other sequences of positive real numbers.

**Definition 4**     *Let* $\Lambda : 0 < \lambda_1 \leq \lambda_2 \leq \dots$ *be an increasing sequence of positive real numbers. The zeta-function attached to this sequence is defined by*

$$\zeta_\Lambda(s) := \sum_{n \geq 1} \lambda_n^{-s}.$$

We make the following three assumptions about $\zeta_\Lambda(s)$:

Z1:  $\zeta_\Lambda(s)$ converges for Re $s \gg 0$;

Z2:  $\zeta_\Lambda(s)$ has a meromorphic continuation to the whole complex plane;

Z3:  $\zeta_\Lambda(s)$ has no pole at $s = 0$.

With these assumptions, we define

**Definition 5**    $\lambda_1 \cdot \lambda_2 \cdot \lambda_3 \cdot \ldots := e^{-\zeta_\Lambda'(0)}$.

Notice the following two lemmas:

**Lemma 1**    *For any integer $N \geq 1$, we have*
$$\lambda_1 \cdot \lambda_2 \cdot \lambda_3 \cdot \ldots = (\lambda_1 \cdot \lambda_2 \cdot \ldots \cdot \lambda_N) \cdot (\lambda_{N+1} \cdot \lambda_{N+2} \cdot \ldots).$$

*Proof.* We have
$$\zeta_\Lambda(s) = \sum_{n=1}^{N} \lambda_n^{-s} + \zeta_N(s),$$
where $\zeta_N(s) = \sum_{n \geq N+1} \lambda_n^{-s}$. Differentiating and putting $s = 0$, we get
$$\zeta_\Lambda'(0) = -\sum_{n=1}^{N} \log \lambda_n + \zeta_N'(0).$$
Now the lemma follows by taking the exponential on both sides.    □

**Lemma 2**    *For any positive real number $a$, we have*
$$(a\lambda_1) \cdot (a\lambda_2) \cdot (a\lambda_3) \cdot \ldots = (\lambda_1 \cdot \lambda_2 \cdot \lambda_3 \cdot \ldots) a^{\zeta_\Lambda(0)}.$$

*Proof.* Clearly
$$\zeta_{a\Lambda}(s) = a^{-s} \zeta_\Lambda(s).$$
Differentiating, putting $s = 0$ and taking the exponential now implies the lemma.    □

**Definition 6**    *Let $\Lambda : 0 < \lambda_1 \leq \lambda_2 \leq \ldots$ be an increasing sequence of positive real numbers. The theta-function attached to this sequence is defined by*
$$\Theta_\Lambda(t) := \sum_{n \geq 1} e^{-\lambda_n t}.$$

We make the following two assumptions about $\Theta_\Lambda(t)$:

$\Theta 1$:  $\Theta_\Lambda(t)$ converges for $t > 0$.

$\Theta 2$:  For every $k \in \mathbf{N}$, there are real numbers $a_i$ $(i \in \mathbf{Z})$, with $a_i = 0$ for $i < -d$, such that
$$\Theta_\Lambda(t) = \sum_{n=-d}^{k} a_n t^n + O(t^{k+1}) \quad \text{for } t \to 0.$$

We note that $\Theta 1$ implies that

$\Theta 3$:  $\Theta_\Lambda(t)$ is $O(e^{-ct})$ for $t \to \infty$, for some positive real number $c$.

Indeed, when $t > 1$

$$\Theta_\Lambda(t) = \sum_{n\geq 1} e^{-\lambda_n(t-1)-\lambda_n} \leq e^{-\lambda_1(t-1)}\Theta_\Lambda(1).$$

We now have the following:

**Theorem 1**   *If $\Theta_\Lambda(t)$ satisfies $\Theta 1$ and $\Theta 2$, then*

$$\zeta_\Lambda(s) = \frac{1}{\Gamma(s)} \int_0^\infty \Theta_\Lambda(t) t^{s-1} dt$$

*and this function satisfies Z1, Z2, Z3. Furthermore, we have*

$$\zeta_\Lambda(0) = a_0,$$

*and*

$$\zeta_\Lambda'(0) = \text{finite part of} \left( \int_\epsilon^\infty \Theta_\Lambda(t)\frac{dt}{t} \right)_{\epsilon\to 0} + \gamma a_0.$$

*Proof.* Because of $\Theta 2$ and $\Theta 3$ the integral defining $\zeta_\Lambda(s)$ converges for $\text{Re } s > d$, hence Z1 is satisfied. By $\Theta 2$ we can write, for every $k \in \mathbf{N}$,

$$\Theta_\Lambda(t) = \sum_{n\leq k} a_n t^n + \rho_k(t)$$

with $\rho_k(t) = O(t^{k+1})$. Therefore
(10)
$$\zeta_\Lambda(s) = \frac{1}{\Gamma(s)} \int_1^\infty \Theta_\Lambda(t)t^{s-1}dt + \sum_{n\leq k} \frac{a_n}{\Gamma(s)} \int_0^1 t^{n+s-1}dt$$

$$+ \frac{1}{\Gamma(s)} \int_0^1 \rho_k(t)t^{s-1}dt$$

$$= \frac{1}{\Gamma(s)} \int_1^\infty \Theta_\Lambda(t)t^{s-1}dt + \sum_{n\leq k} \frac{a_n}{\Gamma(s)(n+s)} + \frac{1}{\Gamma(s)} \int_0^1 \rho_k(t)t^{s-1}dt.$$

In the last expression, the first integral is holomorphic for all $s \in \mathbf{C}$, while the second integral is holomorphic for $\text{Re } s > -k - 1$; the sum over $n$ is a meromorphic function in the whole complex plane. As $k$ is arbitrary, we obtain a meromorphic continuation of $\zeta_\Lambda(s)$ to the whole complex plane, which establishes Z2.

Putting $k = 0$ and $s = 0$, we obtain from (10)

$$\zeta_\Lambda(0) = \left( \frac{a_0}{\Gamma(s)\cdot s} \right)_{s=0} + \sum_{n<0} \left( \frac{a_n}{\Gamma(s)(n+s)} \right)_{s=0} = a_0,$$

which proves Z3. Furthermore, by differentiating (10), we get

$$\zeta_\Lambda'(0) = \int_1^\infty \Theta_\Lambda(t) \left(\frac{t^s}{\Gamma(s)}\right)'_{s=0} \frac{dt}{t} + \sum_{n\leq 0} \left(\frac{a_n}{\Gamma(s)(n+s)}\right)'_{s=0}$$

$$+ \int_0^1 \rho_0(t) \left(\frac{t^s}{\Gamma(s)}\right)'_{s=0} \frac{dt}{t}.$$

But since

$$(t^s/\Gamma(s))'_{s=0} = 1,$$

$$(1/\Gamma(s) \cdot s)'_{s=0} = (1/\Gamma(s+1))'_{s=0} = -\Gamma'(1)/\Gamma(1)^2 = \gamma,$$

and

$$\Big(1/(\Gamma(s)(n+s))\Big)'_{s=0} = 1/n,$$

we obtain

$$(11) \qquad \zeta_\Lambda'(0) = \int_1^\infty \Theta_\Lambda(t) \frac{dt}{t} + \gamma a_0 + \sum_{n<0} \frac{a_n}{n} + \int_0^1 \rho_0(t) \frac{dt}{t}.$$

On the other hand, we have for $0 < \epsilon < 1$

$$\int_\epsilon^\infty \Theta_\Lambda(t) \frac{dt}{t} = \int_1^\infty \Theta_\Lambda(t) \frac{dt}{t} + \sum_{n\leq 0} a_n \int_\epsilon^1 t^{n-1} dt + \int_\epsilon^1 \rho_0(t) \frac{dt}{t}$$

$$(12) \qquad = \int_1^\infty \Theta_\Lambda(t) \frac{dt}{t} - a_0 \log \epsilon + \sum_{n<0} a_n \left(\frac{1}{n} - \frac{\epsilon^n}{n}\right)$$

$$+ \int_0^1 \rho_0(t) \frac{dt}{t} + O(\epsilon).$$

The finite part of (12) as $\epsilon \to 0$ is defined to be its limit once the divergent summands $-a_0 \log \epsilon$ and $-a_n \frac{\epsilon^n}{n}$, $n < 0$, have been removed. Comparing with (11) we get

$$\zeta_\Lambda'(0) = \text{finite part of} \left(\int_\epsilon^\infty \Theta_\Lambda(t) \frac{dt}{t}\right)_{\epsilon\to 0} + \gamma a_0,$$

as claimed. □

## 1.3 Questions and complements

**1.3.1**   Physicists have used other ways of regularizing infinite products known as Pauli–Villars regularization and dimensional regularization (see for instance [R]). For which class of positive numbers do they make sense? Do they lead to other justifications of formula (1)?

**1.3.2**   Given two sequences

$$\Lambda : 0 < \lambda_1 \leq \lambda_2 \leq \dots$$

and
$$\mu : 0 < \mu_1 \leq \mu_2 \leq \dots,$$
can one compare
$$(\lambda_1\lambda_2\lambda_3 \dots)(\mu_1\mu_2\mu_3 \dots)$$
with
$$(\lambda_1\mu_1)(\lambda_2\mu_2)(\lambda_3\mu_3)\dots?$$
See [K], 6.5.

**1.3.3**   If $\Lambda$ satisfies $\Theta 1$ and $\Theta 2$, and if $\lambda$ is a positive real number, the sequence
$$0 < \lambda_1 + \lambda \leq \lambda_2 + \lambda \leq \dots$$
also satisfies these properties. So we may consider the regularized product
$$D(\lambda) = (\lambda_1 + \lambda)(\lambda_2 + \lambda)\dots.$$
This function $D(\lambda)$ can in fact, using Lemma 1, be defined for any complex number $\lambda$ and, according to Voros [Vo] and Cartier–Voros [CV], it can be characterized as follows. It is the unique holomorphic function of the complex variable $\lambda$ whose zeroes (counted with multiplicity) are the numbers $-\lambda_1, -\lambda_2, \dots$, which is bounded by $\exp(a + b|\lambda|^N)$ for some constants $a, b$ and $N$, and which admits an asymptotic development of the form
$$\log D(\lambda) = \sum_{k=-d}^{k=m-1} a_k\phi_k(\lambda) + O(\lambda^{-m})$$
for all $m \geq 0$ when $\lambda$ is a positive real number going to infinity. Here $\phi_k(\lambda)$, $k \in \mathbf{Z}$, is the sequence of functions defined by
$$\phi_k(\lambda) = \lambda^{-k}, k \geq 1,$$
$$\phi_0(\lambda) = \log(\lambda),$$
and
$$\phi_{-k}(\lambda) = (\log(\lambda) - (1 + \frac{1}{2} + \dots + \frac{1}{k}))\frac{\lambda^k}{k!}, k \geq 1.$$

**1.3.4**   Regularization of infinite products can be used in the study of zeta functions of algebraic varieties over number fields. In [De1] and [De2], Deninger shows that every local factor of these zeta functions is the regularized characteristic power series of some operator in an infinite-dimensional cohomology theory.

One may also ask whether products on all prime integers (Euler products) can be regularized. A naive question along these lines is: for which values of $s$ can one make sense of a formula such as

$$\zeta(s) = (\prod_p p^s)/(\prod_p (p^s - 1)),$$

where both the numerator and the denominator are regularized products? But, according to a general result of Dahlquist ([Da], see also [Ku]), Euler products of the form $\prod_p h(p^{-s})$ have a natural boundary on the line $\text{Re}(s) = 0$, except for very few holomorphic functions $h(z)$. By taking the logarithm of this statement, we see that Poincaré series over primes are in general not defined at zero, and this forbids us to zeta regularize Euler products. For instance $\sum_p p^{-s}$ has a natural boundary on $\text{Re}(s) = 0$ [LW], hence the product of all primes does not make sense by this method. Is there another way of defining it?

## 2. Generalized Laplacians

**2.1**    Let $M$ be a smooth real manifold and $E$ a smooth vector bundle on $M$. Choose any connection $\nabla$ on $E$ (see §IV.2.1). Given any vector field $X$ on $M$, we get from $\nabla$ a covariant derivative $\nabla_X$ on $\Gamma(M, \text{End}(E))$ by evaluating differentials on $X$.

**Definition 7**    *The algebra of differential operators on $E$, denoted $D(M, E)$, is the subalgebra of $\text{End}_{\mathbb{C}}(\Gamma(M, E))$ generated by $\Gamma(M, \text{End}(E))$ and the $\nabla_X$, for all vector fields $X$.*

If $\nabla'$ is any other connection on $E$, and $X$ any vector field on $M$, then $\nabla'_X - \nabla_X$ lies in $\Gamma(M, \text{End}(E))$, so $D(M, E)$ is in fact independent of our choice of $\nabla$. If $H \in D(M, E)$, we will say that it is of order less or equal to $n$ if it lies in the subspace generated by $\Gamma(M, \text{End}(E))$ and products of at most $n$ covariant derivatives.

Let $U$ be an open set of $M$ contained in a coordinate patch $(x^1, \ldots, x^m)$ $(m = \dim X)$, so that $E$ may be trivialized over it. Then, on $\Gamma(U, E)$, such an $H$ may be (uniquely) written as

$$\sum_{j=0}^n \sum_{a_1 + \cdots + a_m = j} \varphi_a \partial_1^{a_1} \cdots \partial_m^{a_m}$$

where the $\varphi_a$'s are sections of $\text{End}(E)$ over $U$, and $\partial_i$ stands for $\frac{\partial}{\partial x^i}$.

We now fix a Riemannian metric $g$ on $M$.

**Definition 8**    *A generalized Laplacian on $E$ (with respect to $g$) is a differential operator on $E$ whose expression in any local coordinates is the sum of $-\sum_{i,j} g^{ij} \partial_i \partial_j$ with an operator of order $\leq 1$.*

In this definition $g^{ij}$ stands for $g(dx^i, dx^j)Id_E$.

**Lemma 3**    *A differential operator $H$ of order $\leq 2$ is a generalized Laplacian if and only if, for any $f \in C^\infty(M)$ we have the commutation relation in $\mathrm{End}_{\mathbb{C}}(\Gamma(M, E))$ : $[[H, f], f] = -2|df|^2$.*

This lemma provides us with a coordinate-free characterization of generalized Laplacians.

*Proof.* Choose coordinates $(x^i, 1 \leq i \leq m)$ on an open set $U$, and a trivialization of $E$ on $U$. We denote by $[.,.]$ the commutator of two linear endomorphisms of $\Gamma(M, E)$, and we identify any function $f$ with the operator of multiplication by $f$.

Using the basic fact $[\partial_i, f] = \partial_i f$ ( $:= \frac{\partial f}{\partial x^i}$) we compute:

$$[[\partial_i, f], f] = 0$$
$$[\partial_i \partial_j, f] = [\partial_i, f]\partial_j + \partial_i[\partial_j, f] = (\partial_i f)\partial_j + (\partial_i \partial_j f) + (\partial_j f)\partial_i$$
$$[[\partial_i \partial_j, f], f] = 2(\partial_i f)(\partial_j f).$$

From this the claim follows.    □

**2.2**    Now we shall give an important example of generalized Laplacians. Let $X$ be a Kähler manifold and $E$ an hermitian holomorphic vector bundle on $X$. The Kähler metric on $X$ is an hermitian metric $h$ on the holomorphic tangent space of $X$, i.e. the subbundle $T^{1,0}X$ of the complex tangent bundle of $X$. This metric induces a metric on the dual of $T^{1,0}X$, i.e. differential forms of type $(1,0)$, and, by complex conjugation, on forms of type $(0,1)$. By taking the exterior powers of this metric, and by tensoring with the hermitian metric on $E$ we get a pointwise scalar hermitian product $\langle s(x), t(x)\rangle$ for two sections of $A^{0,q}(X, E) = A^{0,q}(X) \otimes_{C^\infty(M)} A^0(X, E)$. On the other hand, let $\omega_0$ be the normalized Kähler form, given in any local chart $(z_\alpha)$ on $X$ by

$$\omega_0 = \frac{i}{2\pi} \sum_{\alpha, \beta} h\left(\frac{\partial}{\partial z_\alpha}, \frac{\partial}{\partial z_\beta}\right) dz_\alpha d\bar{z}_\beta.$$

The $L^2$-scalar product of two sections $s, t \in A^{0,q}(X, E)$ is defined by the formula

$$\langle s, t\rangle_{L^2} = \int_X \langle s(x), t(x)\rangle \frac{\omega_0^n}{n!},$$

where $n = \dim_{\mathbb{C}} X$.

By the Leibniz rule we extend the Cauchy–Riemann operator $\bar{\partial} = \bar{\partial}_E$ on $E$ (see §IV.2.2) to forms of type $(0, q)$ with values in $E$. We obtain this way the Dolbeault complex

$$A^{0,0}(X, E) \xrightarrow{\bar{\partial}} A^{0,1}(X, E) \xrightarrow{\bar{\partial}} \cdots \xrightarrow{\bar{\partial}} A^{0,q}(X, E) \xrightarrow{\bar{\partial}} \cdots$$

whose cohomology is known to be the sheaf cohomology of $X$ with co-efficients in $E$ [GH].

The operator $\overline{\partial}$ has an adjoint for the $L^2$-scalar product, i.e. there is a map

$$\overline{\partial}^* : A^{0,q+1}(X, E) \to A^{0,q}(X, E)$$

such that

$$\langle s, \overline{\partial}^* t \rangle_{L^2} = \langle \overline{\partial} s, t \rangle_{L^2}$$

for any $s \in A^{0,q}(X, E)$ and $t \in A^{0,q+1}(X, E)$. For any integer $q \geq 0$, the operator $\Delta^q = \overline{\partial}\,\overline{\partial}^* + \overline{\partial}^*\overline{\partial}$ on $A^{0,q}(X, E)$ will be called the *Laplace operator*.

**Lemma 4**     *The operator $2\Delta^q$ is a generalized Laplacian in the sense of Definition 8.*

*Proof.* Let $D := \sqrt{2}(\overline{\partial} + \overline{\partial}^*)$ acting upon $\bigoplus_{q\geq 0} A^{0,q}(X, E)$. Clearly $D^2 = 2\Delta^q$ on $A^{0,q}$. We shall give another description of $D$, known as the *Dirac operator*.

We define the *Clifford action* of $A^1(X)$ on $\bigoplus_{q\geq 0} A^{0,q}(X, E)$ as follows. Given $s \in A^{0,q}(E)$ and $y \in A^1(X)$, with $y = y' + y''$, $y' \in A^{1,0}(X)$ and $y'' \in A^{0,1}(X)$, then:

$$c(y)s = \sqrt{2}(y'' \wedge s - i_{y'}(s)).$$

Here $i_{y'}$ denotes the contraction with the tangent vector of type $(0,1)$ corresponding to $y'$ under the isomorphism induced by the metric on $X$, as above. In other words, $i_{y'}$ is characterized by the equality

$$\langle i_{y'}(s), t \rangle = \langle s, \overline{y}' \wedge t \rangle$$

where $\langle \alpha \otimes u, \beta \otimes v \rangle$ is the function $\langle \alpha, \beta \rangle_X \langle u, v \rangle_E$, for any $\alpha, \beta \in A^{0,q}(X)$, and $u, v \in \Gamma(E)$. We compute

$$c(y)^2 s = 2(y'' \wedge y'' \wedge s - y'' \wedge i_{y'}(s) - i_{y'}(y'' \wedge s) + i_{y'} i_{y'}(s))$$
$$= 2(-y'' \wedge i_{y'}(s) - (\langle y', \overline{y}'' \rangle s - y'' \wedge i_{y'}(s)))$$
$$= -2\langle y', \overline{y}'' \rangle s.$$

In particular, if $y$ is real $(y' = \overline{y}'')$, then $c(y)^2 = -2|y|^2$.

Let

$$\nabla : \bigoplus_{q\geq 0} A^{0,q}(E) \to A^1(X) \otimes \left( \bigoplus_{q\geq 0} A^{0,q}(E) \right)$$

be the tensor product of the Levi–Civita connection of $X$ with the hermitian holomorphic connection of $E$ (see Lemma IV.1). Then the following holds on $\bigoplus_{q\geq 0} A^{0,q}(E)$

$$D = c \circ \nabla.$$

We shall not prove this fact (see for example [BGV], Proposition 3.72). Notice however that $\sqrt{2}\,\bar{\partial} = c \circ \nabla^{0,1}$ is clear from the definitions. The identity $\sqrt{2}\,\bar{\partial}^* = c \circ \nabla^{1,0}$ can be proved up to lower order terms, which is enough for our purpose, by using the fact that $\nabla$ is unitary, since the metric on $X$ is Kähler, and that taking the symbol of an operator commutes with adjunction.

As $\nabla$ is a connection $[\nabla, f] = df$ for $f \in C^\infty(M)$ so we obtain

$$[D, f] = c(df).$$

So

$$[D^2, f] = c(df)D + Dc(df),$$

and

$$[[D^2, f], f] = c(df)c(df) + c(df)c(df) = 2c(df)^2.$$

Finally, using the facts that for $f$ real $c(df)^2 = -2|df|^2$ and that $D^2 = 2\Delta^q$ on $A^{0,q}(E)$ we obtain

$$[[2\Delta^q, f], f] = -2|df|^2.$$

By Lemma 3, and the fact that the order of $\Delta^q$ is less or equal to 2 (since $D$ has order one), we conclude that $2\Delta^q$ is a generalized Laplacian.

$\square$

## 3. Heat kernels

**3.1**   Let $M$ be a Riemannian manifold with metric $g$, $E$ a $C^\infty$-vector bundle on $M$ and $H$ a generalized Laplacian acting on sections of $E$. We denote by $|\Lambda|^{1/2}$ the bundle of half-densities on $M$, defined as in [BGV] before Proposition 1.20.

**Definition 9**   *A heat kernel for $H$ is a family of sections*

$$p_t(x, y) \in (E \otimes |\Lambda|^{1/2})_x \otimes (E^* \otimes |\Lambda|^{1/2})_y$$

*depending on $t \in \mathbb{R}_+^*$ such that the following conditions hold:*

*(i)*   $p_t(x, y)$ *is $C^\infty$ in* $(t, x, y) \in \mathbb{R}_+^* \times M \times M$ ;

*(ii)*   *for every $y$ we have* $(\partial_t + H_x)p_t(x, y) = 0$ ;

*(iii)*   *for any continuous section $s$ with compact support of $E \otimes |\Lambda|^{1/2}$*

$$\lim_{t \to 0} \int_{y \in M} p_t(x, y)s(y) = s(x)$$

*the limit being for the sup-norm, after any choice of a metric on $E$.*

A heat kernel $p_t(x, y)$ defines operators

$$P_t : \Gamma_c(E \otimes |\Lambda|^{1/2}) \longrightarrow \Gamma(E \otimes |\Lambda|^{1/2})$$

by

$$P_t s(x) = \int_M p_t(x, y) s(y).$$

Condition (ii) may be rewritten as

(ii')                         $$(\partial_t + H) P_t = 0.$$

This is the *heat equation* associated to $H$. The condition (iii) is an initial condition for the first order differential equation (ii'), namely: $\lim_{t \to 0} P_t = \mathrm{Id}$. So we should think of $P_t$ as being $e^{-tH}$.

We are now able to state the main result of this section:

**Theorem 2**    *Assume $M$ is compact. Then any generalized Laplacian on $M$ has a unique heat kernel.*

**3.2**    The proof of Theorem 2 is divided into five steps:

I:     uniqueness assuming the existence of $(C^2)$ heat kernels;
II:    existence of a heat kernel for $M = \mathbf{R}^n$ and $H = -\sum_i \partial_i^2$;
III:   construction of a formal solution;
IV:    construction of an approximate solution;
V:     construction of an exact solution from an approximate one by a perturbation process.

**Step I**    We have a pairing $\langle \, , \, \rangle$ between sections of $E$ and sections of $E^* \otimes |\Lambda|$ defined by

$$\langle s, u \rangle = \int_M \langle s(x), u(x) \rangle_x$$

where $\langle \, , \, \rangle_x$ is the natural pairing

$$E_x \times (E_x^* \otimes |\Lambda|_x) \longrightarrow |\Lambda|_x.$$

This enables us to associate to $H$ an adjoint operator $H^*$, acting on sections of $E^* \otimes |\Lambda|$. It is characterized by the equality $\langle Hs, u \rangle = \langle s, H^*u \rangle$, for all $s$ and $u$. As it turns out by an easy computation, $H^*$ is a generalized Laplacian on $E^* \otimes |\Lambda|$. So let us assume that there exists a heat kernel $P_t$ for $H$ and $P_t^*$ for $H^*$. Let $s$ be a section of $E$, and $u$ a section of $E^* \otimes |\Lambda|$. Fix $t > 0$ and let $f(\theta) = \langle P_\theta s, P_{t-\theta}^* u \rangle$ for $0 < \theta < t$. By (ii') we have

$$f'(\theta) = \langle -HP_\theta s, \ P_{t-\theta}^* u \rangle + \langle P_\theta s, H^* P_{t-\theta}^* u \rangle = 0.$$

By (iii)

$$\lim_{\theta \to 0} f(\theta) = \langle s, P_t^* u \rangle,$$

$$\lim_{\theta \to t} f(\theta) = \langle P_t s, u \rangle.$$

Therefore $\langle P_t s, u \rangle = \langle s, P_t^* u \rangle$. This shows that $P_t$, hence $p_t(x,y)$, is determined by $P_t^*$. So the uniqueness of the heat kernel is proved.   □

**Step II**   Assume $M = \mathbb{R}^n$, $E$ is the trivial line bundle $\mathbb{R}^n \times \mathbb{C}$, and $H = \Delta$ is the standard Laplacian $-\sum_i \partial_i^2$. The heat kernel can then be given by an explicit formula:

(13)        $p_t(x,y) = (4\pi t)^{-n/2} e^{-|x-y|^2/4t} |dx|^{1/2} \otimes |dy|^{1/2}.$

Condition (i) of Definition 9 is clearly satisfied. As for (ii), it is sufficient to check it when $y = 0$, since both sides are invariant under translation. Now

$$p_t(x,0) = \prod_{i=1}^{n} [(4\pi t)^{-1/2} e^{-x_i^2/4t} |dx_i|^{1/2} \otimes |dy_i|^{1/2}].$$

So it is sufficient to check that

$$(\partial_t - \partial_x^2)[(4\pi t)^{-1/2} e^{-x^2/4t}] = 0.$$

This is an easy computation:

$$\partial_t[(4\pi t)^{-1/2} e^{-x^2/4t}]$$

$$= -\frac{1}{2}(4\pi)^{-1/2} t^{-3/2} e^{-x^2/4t} + \frac{1}{4}(4\pi t)^{-1/2} t^{-2} x^2 e^{-x^2/4t},$$

$$\partial_t[(4\pi t)^{-1/2} e^{-x^2/4t}] = (4\pi t)^{-1/2}(-\frac{1}{2t})[e^{-x^2/4t} - \frac{x^2}{2t} e^{-x^2/4t}].$$

Finally, (iii) can be reduced to a well-known one-dimensional identity:

$$\lim_{t \to 0} \int_{\mathbb{R}} \frac{1}{\sqrt{4\pi t}} e^{-\frac{(x-y)^2}{4t}} s(y) dy = s(x)$$

where $s$ is a continuous function on $\mathbb{R}$ with compact support.   □

### 3.3

**Step III**   We begin by defining *normal coordinates*. Recall that a smooth path $x : [0,1] \to M$ is said to be a *geodesic* if it minimizes the functional $L(x) = \int_0^1 |\dot{x}(t)| dt$. This leads to an Euler–Lagrange equation, which is of order 2. From this follows that for any $y \in M$, and any $v \in T_y M$ small enough, there exists a unique geodesic $x : [0,1] \to M$ with initial conditions $x(0) = y$, $\dot{x}(0) = v$. This enables us to define $\exp v := x(1)$. For a small enough $\epsilon > 0$, this exponential gives a diffeomorphism

$$\exp : \begin{pmatrix} \text{ball in } T_y M \\ |v| < \epsilon \end{pmatrix} \xrightarrow{\sim} \begin{pmatrix} \text{open neighborhood} \\ \text{of } y \text{ in } M \end{pmatrix}.$$

Identifying $T_yM$ with $\mathbb{R}^n$ we get coordinates near $y$. They are called *normal coordinates*, and in the following they will be used to identify any $x \in M$ close to $y$ with a point in $\mathbb{R}^n$, also denoted $x$.

We now look for a solution of the heat equation by a *formal* power series of the type :

$$k_t(x,y) = q_t(x,y) \left( \sum_{i \geq 0} \phi_i(x,y,H)t^i \right) |dy|^{1/2},$$

where the coefficients $\phi_i(x,y,H) \in E_x \otimes E_y^*$ are smooth sections defined in a neighborhood of the diagonal in $M \times M$ and

$$q_t(x,y) = (4\pi t)^{-n/2} e^{-|x-y|^2/4t} |dx|^{1/2}$$

as in (13). One can check the following; see [BGV], Proposition 2.24.

**Lemma 5** *For any family of sections $s_t$ of $\Gamma(E)$, defined in a neighborhood of $y$ and smooth in $t$,*

$$(\partial_t + H)(q_t s_t) = q_t(\partial_t + t^{-1}\nabla_{\mathcal{R}} + j^{1/2}H\ j^{-1/2})(s_t),$$

*where $\mathcal{R} = \sum_{i=1}^n x^i \partial_i$ is the radial vector field and*

$$j(x) = |\det g_{ij}(x)|^{1/2}.$$

Now let $B = j^{1/2}H\ j^{-1/2}$.

**Proposition 3** *There exist unique sections $\phi_i(x,y,H)$ of $E_x \otimes E_y^*$, for all $i \geq 0$, such that:*
(a) $\phi_0(y,y,H) = \mathrm{id}_{E_y}$;
(b) $(\partial_t + t^{-1}\nabla_{\mathcal{R}} + B_x)(\sum_{i \geq 0} \phi_i t^i) = 0.$

*Proof.* The equality (b) of the formal power series in $t$ is equivalent to the equalities

$$\nabla_{\mathcal{R}}\phi_0 = 0 \qquad \text{if} \quad i = 0;$$
$$(\nabla_{\mathcal{R}} + i)\phi_i = -B_x\phi_{i-1} \qquad \text{if} \quad i > 0.$$

Introducing $f_i(s) = s^i\phi_i(\exp(sx),y,H)$, the conditions become, when $x,y$ are fixed,

$$f_0(0) = \mathrm{id}_{E_y}, \ f_i(0) = 0 \quad \text{if} \quad i > 0\ ,$$

and

$$\frac{d}{ds}f_i(s) = (i + \nabla_{\mathcal{R}})s^{i-1}\phi_i(\exp(sx),y,H),$$

i.e.

$$\frac{d}{ds}\ f_0(s) = 0$$

and

$$\frac{d}{ds}\, f_i(s) = -s^{i-1} B_x \phi_{i-1} \quad \text{when} \quad i > 0.$$

We have reduced the problem to differential equations of order one. By induction on $i$ this shows the existence of $\phi_i(x, y, H) = f_i(1)$.

**Remark**     The proof shows that there exist *universal* and *local* formulas for $\phi_i(x, x, H)$ in terms of the derivatives of the $g^{ij}$ and of the coefficients of $H$; but these are difficult to make explicit, see [Gi]. In particular, if $M, E, H$ vary smoothly with respect to a parameter $b$, then the same is true for the $\phi_i$'s.                                        □

### 3.4

**Step IV**     Since $M$ is compact, there exists an $\epsilon > 0$ such that exp is well defined on $\{v \in T_y M : |v| < \epsilon\}$ for any $y$. Choose a smooth function $\chi : \mathbb{R}_+ \to [0, 1]$ satisfying

$$\chi(s) = 1 \quad \text{if } s < \epsilon^2/4,$$
$$\chi(s) = 0 \quad \text{if } s > \epsilon^2,$$

and let $N \in \mathbb{N}$ be an integer. Define

$$(14) \qquad k_t^N(x, y) := \chi(d(x,y)^2) q_t(x,y) \left( \sum_{i=0}^N t^i \phi_i(x,y,H) \right) |dy|^{1/2};$$

in this expression $d(x, y)$ is the infimum of the lengths of all paths from $x$ to $y$. Clearly $k_t^N(x, y)$ depends smoothly on $t, x$ and $y$. So we may view it as the kernel of an operator $K_t^N$ acting on $\Gamma(M, E \otimes |\Lambda|^{1/2})$. In the following proposition $n = \dim M$, $\alpha \in \mathbb{N}$, and $\| \cdot \|_\alpha$ is a norm involving at most $\alpha$ derivatives with respect to $x, y$.

### Proposition 4

(i)     *For all $T > 0$, $K_t^N$ is uniformly bounded for $\| \cdot \|_\alpha$ in the range $0 \le t \le T$.*

(ii)     *For any section $s$ of $E \otimes |\Lambda|^{1/2}$*

$$\lim_{t \to 0} K_t^N \cdot s = s \qquad \text{for the norm} \qquad \| \cdot \|_\alpha.$$

(iii)     *As $t \to 0$ we have estimates:*

$$\|(\partial_t + H_x) k_t^N(x, y)\|_\alpha \le O\left(t^{N - \frac{n}{2} - \frac{\alpha}{2}}\right).$$

This means that the $k_t^N$ give an approximate solution to the heat equation. For a proof, see [BGV], Thm. 2.29.

**3.5**

**Step V**    In the sequel, we fix an $N \geq n/2$, and drop the $N$ in our notation. To motivate what follows, we first assume that $M$ is a point. We then have a vector space $V$, a linear operator $H$ on $V$, and a family $K_t$ of endomorphisms of $V$ such that $K_0 = \text{id}$, $\|\partial_t K_t + H\, K_t\| = O(t^\beta)$ as $t \to 0$, and $K_t$ is uniformly bounded when $0 \leq t \leq T$.

Let $R_t = \partial_t K_t + H\, K_t$. Let us consider the simplex

$$\Delta_t^k = \{(t_1, \dots, t_k) \in \mathbf{R}^k | 0 \leq t_1 \leq t_2 \leq \dots, \leq t_k \leq t\},$$

and define

$$Q_t^k = \int_{\Delta_t^k} K_{t-t_k} R_{t_k - t_{k-1}} \cdots R_{t_2 - t_1} R_{t_1} dt_1 \dots dt_k.$$

**Proposition 5**    *The series $\sum_{k \geq 0} (-1)^k Q_t^k$ converges absolutely. Its sum $P_t$ is such that $P_0 = \text{id}$, $\partial_t P_t + H P_t = 0$, and $\|P_t - K_t\| = O(t^{1+\beta})$.*

*Proof.*  Using $\text{vol}(\Delta_t^k) = t^k/k!$ we see that $\|Q_t^k\| = O(t^k/k!)$, and this implies the convergence of the series. Now recall the following differentiation rule:

$$\partial_t \left( \int_0^t f(t-s)g(s)ds \right) = \int_0^t \frac{\partial}{\partial t} f(t-s)g(s)ds + f(0)g(t).$$

Take $f(s) = K_s$ and

$$g(s) = R^{k-1}(s) = \int_{\Delta_s^{k-1}} R_{s-t_{k-1}} \dots R_{t_1} dt_1 \dots dt_{k-1}.$$

This gives $(\partial_t + H)Q_t^k = R^k(t) + R^{k-1}(t)$. Using the definition of $P_t$ as an alternating sum we get $(\partial_t + H)P_t = 0$. Since $Q_t^0 = K_t$, the last assertion is also clear.  $\square$

When $M$ is arbitrary we follow essentially the same pattern, the convolution of kernels taking the place of the multiplication of operators. Define

$$r_t(x,y) = (\partial_t + H_x)k_t(x,y),$$

$$q_t^k(x,y) = \int_{\Delta_t^k} \int_{M^k} k_{t-t_k}(x, z_k) r_{t_k - t_{k-1}}(z_k, z_{k-1}) \cdots r_{t_1}(z_1, y) dt_1 \cdots dt_k,$$

and

$$r_t^k(x,y) = \int_{\Delta_t^{k-1}} \int_{M^{k-1}} r_{t-t_{k-1}}(x, z_{k-1}) \cdots \cdots \cdots r_{t_1}(z_1, y) dt_1 \cdots dt_{k-1}.$$

Suppose that $N \geq \frac{n}{2} + \frac{\alpha}{2}$. Then $q_t^k$ and $r_t^k$ are $C^\alpha$ with respect to $x$ and

$y$ and the following estimates can be derived from Proposition 4:

$$\|q_t^k\|_\alpha \leq A^k \; t^{k(N-n/2)-\alpha/2}\frac{t^k}{(k-1)!}$$

$$\|r_t^k\|_\alpha \leq B^k \; t^{k(N-n/2)-\alpha/2}\frac{t^{k-1}}{(k-1)!}$$

for certain constants $A, B \geq 0$; see [BGV], Lemmas 2.27 and 2.28.

Now essentially the same proof as the one of Proposition 5 gives:

**Theorem 3**     *Let $p_t(x,y) = \sum_{k\geq0}(-1)^k q_t^k(x,y)$. This sum converges absolutely. The form $p_t$ is $C^\alpha$ with respect to $x$ and $y$. It satisfies*

$$(\partial_t + H_x)p_t(x,y) = 0$$

*and $\lim_{t\to0}P_t s = s$ in sup norm. Moreover*

$$\|p_t(x,y) - k_t(x,y)\|_\alpha \leq O(t^{N-n/2-\alpha/2}).$$

We remark that, since it is unique, $p_t$ does not depend on $N$; so it is $C^\infty$ in $(x,y)$. The proof of Theorem 2 is now complete.     □

## 3.6 Complements

**3.6.1**     In the Remark in §3.3 we stated that the formal solution depends smoothly on parameters when this is the case for $M, E, H$. This smoothness is also preserved in Steps IV and V. If $H$ depends smoothly on a parameter $b$, then so does the heat kernel $p_t(x,y)$; see [BGV] Theorems 2.48 and 9.50, and [Gr].

**3.6.2**     A special case of Proposition 5 is the following formula, known as *Duhamel's formula*. Let $V$ be a finite-dimensional vector space over $\mathbb{C}$, and $H = H_0 + H_1$ the sum of two endomorphisms of $V$. Applying Proposition 5 with $K_t = e^{-tH_0}$, hence $R_t = (\partial_t + H)K_t = H_1 e^{-tH_0}$, we obtain

(15)
$$e^{-t(H_0+H_1)} =$$

$$e^{-tH_0} + \sum_{k\geq1}(-1)^k \int_{\Delta_t^k} e^{-(t-t_k)H_0}H_1 e^{-(t_k-t_{k-1})H_0}...H_1 e^{-t_1 H_0}dt_1...dt_k.$$

**3.6.3**     Let $H(\epsilon) \in \mathrm{End}_\mathbb{C}(V)$ be a family of operators depending smoothly on a parameter $\epsilon \in \mathbb{R}$. If we write formula (15) when $t = 1$, $H_0 = H(0)$ and $H_1 = H(\epsilon) - H_0$, we get

$$e^{-H(\epsilon)} = e^{-H_0} - \int_0^1 e^{-(1-s)H_0}(H(\epsilon) - H_0)e^{-sH_0}ds + o(\epsilon).$$

Letting $\epsilon \to 0$ we obtain

$$\text{(16)} \qquad \frac{d}{d\epsilon}e^{-H(\epsilon)}\Big|_{\epsilon=0} = -\int_0^1 e^{-(1-s)H(0)}\left(\frac{d}{d\epsilon}H(\epsilon)\Big|_{\epsilon=0}\right)e^{-sH(0)}ds.$$

**3.6.4** One may also proceed as follows. Let $A$ and $B$ be two endomorphisms of $V$. Then

$$\frac{d}{du}\left(e^{uA}e^{-uB}\right) = e^{uA}(A-B)e^{-uB},$$

so

$$e^A e^{-B} = I + \int_0^1 e^{uA}(A-B)e^{-uB}du,$$

and

$$e^A = e^B + \int_0^1 e^{(1-s)A}(A-B)e^{-sB}ds.$$

This identity enables us to derive both Duhamel's formula (by iteration) and the formula for $\frac{d}{d\epsilon}e^{-H(\epsilon)}$.

**3.6.5** Formulas (15) and (16) remain valid when $H, H_0$ and $H(\epsilon)$ are generalized Laplacians. As in §3.5, they have to be understood as computing the kernels of some operators, and composition of operators means convolutions of kernels; see [BGV], formulas (2.5) and (2.6).

## 4. Infinite determinants

**4.1** Let $\mathcal{H}$ be a (separable) Hilbert space. If $A$ is a compact operator on $\mathcal{H}$, we say that $A$ is *trace-class* if $\sum_{i\geq 0}(Ae_i, e_i)$ is absolutely convergent for one, hence any, orthonormal basis $(e_i)_{i\geq 0}$. The value of this sum, which is independent of the choice of the basis, is called the *trace* of $A$, and denoted $\text{tr}(A)$.

**Lemma 6** *Let $M$ be a compact manifold, $E$ a $C^\infty$ vector bundle on $M$ equipped with a smooth metric, and $A$ an operator on $\Gamma_{L^2}(M, E \otimes |\Lambda|^{1/2})$ such that $A$ acts on smooth sections by convolution with a smooth kernel $a(x,y)$. Then $A$ is trace-class and*

$$\text{tr}(A) = \int_M \text{tr}a(x,x).$$

In this statement, $\Gamma_{L^2}$ is the Hilbert space completion of the space of smooth sections of $E$ for the $L^2$-metric:

$$\langle s, t \rangle = \int_M \langle s(x), t(x) \rangle_x,$$

and $a(x, x)$ is a section of $\mathrm{End}(E) \otimes |\Lambda|$.

*Proof.* We give a quick description; an easier proof for heat kernels may be found in [BGV], Proposition 2.32. Using a (finite) partition of unity subordinate to an open cover of $M$ by small open subsets, we are reduced to the case of a compactly supported kernel over $\mathbb{R}^n$, and the trivial line bundle with standard metric. This may be re-embedded into a torus $T = \mathbb{R}^n / \mathbb{Z}^n$. One then uses the explicit basis of $\Gamma_{L^2}$ given by the Fourier exponentials and one decomposes the kernel as a double Fourier series. The result follows. $\qquad \square$

**4.2**      Suppose now that $H$ is a generalized Laplacian on $M$ acting on sections of $E$. We have by Theorem 2 a corresponding heat kernel $p_t(x, y)$, defining operators $P_t$.

**Definition 10**      *The theta function of $H$ is defined by*

$$\theta(t) = \mathrm{tr}(P_t) = \int_M \mathrm{tr} p_t(x, x)$$

*for all $t > 0$.*

We know that $\theta(t)$ is smooth in $t$ (we integrate over $M$ which is compact). If there are some extra parameters, $\theta(t)$ will also depend smoothly on them by the Remark in 3.3. Using the notation of Theorem 3, we also have, as $t \to 0$

$$\left| \theta(t) - \int_M \mathrm{tr} k_t^N(x, x) \right| = O(t^{N-n/2+1}).$$

Now (14) gives the formula

$$k_t^N(x, x) = (4\pi t)^{-n/2} \left( \sum_{i=0}^{N} \phi_i(x, x, H) t^i \right) |dx|^{1/2} \otimes |dy|^{1/2},$$

where the $\phi_i(x, x, H)$ are given by universal local formulas in terms of the metric of $M$ and the coefficients of $H$. This shows the existence of an asymptotic development

$$\theta(t) = (4\pi t)^{-n/2} \sum_{i=0}^{N} t^i a_i + O(t^{N-n/2+1})$$

as $t \to 0$, for any $N \geq 0$, where

$$a_i = \int_M \mathrm{tr} \phi_i(x, x, H) |dx|$$

are given by universal local formulas. Moreover, if $H$ depends smoothly on some parameter, so do the $a_i$'s.

**4.3**    Assume now that $H = H^*$ on smooth sections. We state without proof Proposition 2.42 of [BGV]

**Lemma 7**    *The closure of $H^*$ on $\Gamma_{L^2}$ is the adjoint of $H$ on $\Gamma_{L^2}$.*

Thus, if $H = H^*$ and $H$ is positive, we get $P_t^* = P_t$ and $H\varphi = \lambda_i \varphi$ if and only if $P_t \varphi = e^{-t\lambda_i}\varphi$ for all $t > 0$. So

$$\theta(t) = \sum_{i \geq 0} e^{-t\lambda_i}$$

for some real numbers $0 < \lambda_1 \leq \lambda_2 \leq \dots$ . From what we have shown, we conclude that $\theta(t)$ converges and that conditions $\Theta 1$ and $\Theta 2$ of §1 are satisfied. So we may introduce the zeta function

$$\zeta_H(s) = \sum_{i \geq 0} \lambda_i^{-s}.$$

Applying Theorem 1 in §1, we know that this series converges when $\operatorname{Re} s > n/2$, that it extends meromorphically to the complex plane, and that it is holomorphic at $s = 0$. The *determinant* of the (generalized) Laplacian $H$ is defined as follows:

$$\det H = e^{-\zeta_H'(0)}.$$

When $H = H^*$ and $H$ is only semi-positive, we consider its restriction to the orthogonal complement of its kernel. We write

$$\zeta_H(s) = \operatorname{tr}'(H^{-s}) , \quad \operatorname{Re} s > n/2 ,$$

for the zeta function of this restriction, and

(17)                     $$\det{}' H = e^{-\zeta_H'(0)}$$

for its determinant.

**4.4**

**Remarks**

**4.4.1**    Few explicit direct computations of determinants of Laplace operators are known; see [RS]. Kronecker's limit formula can be used in the case of a torus [RS]. The case of the trivial line bundle over a projective space is treated in [GS5]. The computation is extended to arbitrary hermitian symmetric spaces in [Ko].

The arithmetic Riemann–Roch–Grothendieck formula in Chapter VIII computes a product of determinants (when the other terms in the formula can be estimated).

**4.4.2**    Sarnak [Sa] and Voros [Vo] have shown that the Selberg zeta function of a compact Riemann surface with its Poincaré metric is equal (up to a the product by a function depending only on the genus) to the regularized characteristic power series of its Laplace operator on functions, defined as in §1.3.3.

For non-compact quotients of the upper-half plane by congruence subgroups, using the Selberg trace formula, Efrat [E1] defined some kind of regularized characteristic power series of the Laplacian, and proved that it is equal to $\varphi(s)L(s)\varphi(2-s)$, where $\varphi(s)$ is a Dirichlet $L$-function [E2] and $L(s)$ has its zeroes on the line $\mathrm{Re}\, s = 1$.

# VI

---

# The Determinant of the Cohomology

Let $f : X \to Y$ be a morphism of arithmetic varieties, smooth on the generic fiber $X_{\mathbb{Q}}$, and $\overline{E}$ an hermitian vector bundle on $X$. We shall define an hermitian line bundle $\lambda(E)_Q = (\lambda(E), h_Q)$ on $Y$.

The algebraic line bundle $\lambda(E)$ is *the determinant of the cohomology*. Its definition is due to Grothendieck and Knudsen–Mumford [KM]. The fiber of $\lambda(E)$ at every point $y$ in $Y$ is the alternated tensor product

$$\lambda(E)_y = \bigotimes_{q \geq 0} (\Lambda^{\max}(H^q(X_y, E)))^{(-1)^q}.$$

In this formula $H^q(X_y, E)$ is the coherent cohomology of the fiber of $X_y = f^{(-1)}(y)$ with coefficients in $E$, $\Lambda^{\max}$ is the maximal exterior power, and $L^{-1}$ denotes the dual of a line bundle $L$. The construction of $\lambda(E)$ is purely algebraic, and given in §1.

On the set of complex points $Y(\mathbb{C})$, the line bundle $\lambda(E)$ induces an holomorphic line bundle. Following Quillen [Q2], once a Kähler metric has been chosen on each fiber of $f$, there is another description of the underlying smooth line bundle $\lambda(E)_\infty$. It uses the family of Laplace operators along the fibers of $f$. This is described in §2 — the properties we need on smooth families of elliptic operators can be found in the book [BGV], Chapter 9; we present here a different approach, based on an earlier draft of it. Following [BGS1], we show in Proposition 2 that this definition is compatible with the algebraic one.

This analytic definition of $\lambda(E)_\infty$ allows one to define the *Quillen metric* $h_Q$ on this line bundle. It is equal to the $L^2$-metric, given by integration on the fibers, multiplied by an alternated product of determinants of Laplace operators (Definition 3), which is also the square of the Ray–Singer analytic torsion [RS]. This definition is due to Quillen in relative dimension one [Q2]. Following [BGS1] we show in §3 that $h_Q$ is smooth (Theorem 3). We also state from [BGS1] a formula for the first Chern form of $\lambda(E)_Q$ (Theorem 4). This Riemann–Roch–Grothendieck theorem at the level of forms will be proved in the next chapter. We explain in Corollary 1 how it leads to an "anomaly" formula saying how $h_Q$ varies with the metric on $E$, and, more generally, what happens for exact sequences of bundles on $X$. From [BGS1] we also state, without proof, how $h_Q$ varies with the metric on the fibers of $f$ (Theorem 5).

## 1. The determinant line bundle: algebraic approach

**1.1**    Let $X$ be a noetherian separated scheme and $E$ a locally free coherent $\mathcal{O}_X$-module. Let $\det^* E = \Lambda^{\max} E$ be the maximal exterior power of $E$. Its fiber at $x \in X$ is $(\det^* E)_x = \Lambda^{\mathrm{rank}(E_x)} E_x$. Given an exact sequence of such $\mathcal{O}_X$-modules:

$$\mathcal{E} : 0 \to S \xrightarrow{\;i\;} E \xrightarrow{\;j\;} Q \to 0$$

there is an associated isomorphism

$$\varphi_{\mathcal{E}} : \det^* S \otimes \det^* Q \xrightarrow{\;\sim\;} \det^* E$$

so that, for local sections $e_1, \ldots, e_n$ of $S$ ($n = \mathrm{rank}\, S$) and $f_1, \ldots, f_m$ of $Q$ ($m = \mathrm{rank}\, Q$),

$$\varphi_{\mathcal{E}}(e_1 \wedge \cdots \wedge e_n \otimes f_1 \wedge \cdots \wedge f_m) = i(e_1) \wedge \cdots \wedge i(e_n) \wedge \widetilde{f_1} \wedge \cdots \wedge \widetilde{f_m}$$

where the $\widetilde{f_i}$'s are liftings of sections of $E$.

Consider now $E_1 = S \oplus Q$, $E_2 = Q \oplus S$, and the two obvious exact sequences

$$\mathcal{E}_1 : 0 \to S \to E_1 \to Q \to 0$$

and

$$\mathcal{E}_2 : 0 \to Q \to E_2 \to S \to 0.$$

Define $T : E_1 \longrightarrow E_2$ by $T(e, f) = (f, e)$ and

$$\tau : \det^* S \otimes \det^* Q \to \det^* Q \otimes \det^* S$$

by $\tau(s \otimes q) = q \otimes s$. Then the composition of the isomorphisms:

$$\det^* S \otimes \det^* Q \xrightarrow[\sim]{\varphi_{\mathcal{E}_1}} \det^* E_1 \xrightarrow[\sim]{\det^* T} \det^* E_2 \xrightarrow[\sim]{\varphi_{\mathcal{E}_2}^{-1}} \det^* Q \otimes \det^* S$$

is *not* $\tau$ but $(-1)^{n \cdot m} \tau$. This is an indication that the line bundles $\det^* S$,

$\det^* Q$ "should not forget" the rank of the original vector bundles, or at least its parity. This is quite natural from the point of view of the filtration of $K$-theory by codimension: the rank $r$ of $E$ is its class in $gr^0 K_0(X)$, while the class of $\det^* E$ is the next invariant, namely the class of $[E] - [\mathcal{O}_X^r]$ in $gr^1 K_0(X)$. This leads to the following definition.

**Definition 1**    *A graded line bundle $(L, \alpha)$ is a pair consisting of a line bundle $L$, and a continuous map $\alpha : X \to \mathbb{Z}/2$. If $E$ is a vector bundle on $X$, the determinant line bundle defined by $E$ is the graded line bundle $\det E = (\det^* E, \operatorname{rank} E)$.*

We can take tensor products:
$$(L, \alpha) \otimes (M, \beta) = (L \otimes M, \alpha + \beta),$$
and we have isomorphisms:
$$(L, \alpha) \otimes (M, \beta) \overset{\tau}{\underset{\sim}{\longrightarrow}} (M, \beta) \otimes (L, \alpha)$$
$$\ell \otimes m \longmapsto (-1)^{\alpha\beta} m \otimes \ell.$$
Given a short exact sequence
$$\mathcal{E} : 0 \to S \to E \to Q \to 0$$
we have $\operatorname{rank} E = \operatorname{rank} S + \operatorname{rank} Q$, and there is an isomorphism $\varphi_{\mathcal{E}} : \det S \otimes \det Q \overset{\sim}{\longrightarrow} \det E$. Now

$$
\begin{array}{ccc}
\det S \otimes \det Q & \overset{\sim}{\longrightarrow} & \det(S \oplus Q) \\
\tau \downarrow & & \downarrow \wr \text{ transposition} \\
\det Q \otimes \det S & \overset{\sim}{\longrightarrow} & \det(Q \oplus S)
\end{array}
$$

is a commutative diagram.

We will also need the definition
$$(L, \alpha)^{-1} = (L^{-1}, \alpha) \qquad [L^{-1} = \operatorname{Hom}(L, \mathcal{O}_X)];$$
recall that $\alpha$ is taken modulo 2. Finally, there is a unit element $1_X = (\mathcal{O}_X, 0)$.

**1.2**    We now assume, moreover, that $X$ is *regular*. Then any coherent $\mathcal{O}_X$-module $M$ has a resolution by locally free coherent $\mathcal{O}_X$-modules:
$$0 \to E_n \to \cdots \to E_0 \to M \to 0.$$
Let us define $\det E_. = \otimes_{i \geq 0} (\det E_i)^{(-1)^i}$. We would like to define $\det M$ to be $\det E_.$. For this we need the following fact:

**Lemma 1**    *Given two complexes $P_.$ and $Q_.$ of vector bundles and a* *quasi-isomorphism* $P_. \xrightarrow{f} Q_.$ *There is an isomorphism*

$$\det P_. \xrightarrow[\sim]{\det f} \det Q_.$$

*which depends only on the homotopy class of* $f$.

Assuming this lemma, suppose we have two resolutions $E_.$ and $F_.$ of $M$ by vector bundles. Locally they are quasi-isomorphic, say by $f : E_. \to F_.$. So we obtain local isomorphisms $\det E_. \xrightarrow{\sim} \det F_.$. If $g : E_. \longrightarrow F_.$ is another local quasi-isomorphism, then on any affine open subset on which both $f$ and $g$ are defined, $f$ and $g$ are homotopic, so they induce the same isomorphism $\det E_. \xrightarrow{\sim} \det F_.$. Therefore there is a well-defined global isomorphism $\det E_. \xrightarrow{\sim} \det F_.$, uniquely characterized by the fact that it is locally induced by $\det f$ for $f : E_. \longrightarrow F_.$ any quasi-isomorphism.

*Proof of Lemma 1.* First we define $\det f : \det P_. \longrightarrow \det Q_.$.

**First Case**    Assume that $Q_. = 0$, hence $P_.$ is acyclic. Locally we may split $0 \to P_n \to \cdots \to P_0 \to 0$ in two exact sequences of vector bundles:

$$R_. : \ 0 \to P_n \to \cdots \to P_2 \to B_1 \to 0,$$
$$S_. : \ 0 \to B_1 \to P_1 \to P_0 \to 0.$$

So, by induction (since $\det P_.$ is clearly isomorphic to $\det R_. \otimes \det S_.$), it is only necessary to define $\det P_. \xrightarrow[\sim]{\det f} 1_X$ when $P_.$ is of length $\leq 2$. We define it as the composition

$$\det P_. = \det P_0 \otimes (\det P_1)^{-1} \otimes \det P_2 \xrightarrow{\sim} \det P_2 \otimes \det P_0 \otimes (\det P_1)^{-1}$$

$$\xleftarrow{\sim} \det P_1 \otimes (\det P_1)^{-1} \xleftarrow{\sim} 1_X.$$

Here, the second isomorphism is induced by $\varphi_{\mathcal{E}}$, where $\mathcal{E}$ is the exact sequence

$$\mathcal{E} : 0 \longrightarrow P_2 \longrightarrow P_1 \longrightarrow P_0 \longrightarrow 0$$

and the first isomorphism is obtained by initially permuting $(\det P_1)^{-1}$ and $\det P_2$, and then $\det P_0$ and $\det P_2$, or directly $\det P_2$ with $\det P_0 \otimes (\det P_1)^{-1}$ (this gives the same sign).

**Second Case**    We suppose that $f$ is split injective:

$$0 \longrightarrow P_. \xrightarrow{f} Q_. \longrightarrow H_. \longrightarrow 0$$

with $H.$ acyclic. Each short exact sequence $0 \to P_n \to Q_n \to H_n \to 0$ gives an isomorphism

$$\det P_n \otimes \det H_n \xrightarrow[\sim]{\varphi_n} \det Q_n.$$

For $n$ odd, we use

$$(\det P_n)^{-1} \otimes (\det H_n)^{-1} \xrightarrow{\sim} (\det H_n)^{-1} \otimes (\det P_n)^{-1}$$

$$\xrightarrow{\sim} (\det H_n \otimes \det P_n)^{-1} \xleftarrow[\sim]{\varphi_n} (\det Q_n)^{-1}.$$

So we have an isomorphism

$$\det P. \otimes \det H. \xrightarrow{\sim} \det Q. .$$

By the first case $\det H. \xrightarrow{\sim} 1_X$. We tensor with $\det P.$ the inverse of this isomorphism, and we obtain an isomorphism $\det f : \det P. \xrightarrow{\sim} \det Q.$.

**Third Case**   Consider now the general case. We use the mapping cylinder $Z.^f$:

$$Z_i^f = P_i \oplus Q_i \oplus P_{i-1}$$

with differentials

$$d_i^{Z^f} = \begin{pmatrix} d_i^P & 0 & -1 \\ 0 & d_i^Q & f_{i+1} \\ 0 & 0 & -d_{i-1}^P \end{pmatrix}.$$

We have maps

$$P. \xrightarrow{\alpha} Z.^f \overset{\beta}{\underset{\gamma}{\rightleftarrows}} Q.$$

defined by

$$\alpha(p) = (p, 0, 0) \qquad\qquad \gamma\beta = \mathrm{id},$$

$$\beta(q) = (0, q, 0) \qquad\qquad \gamma\alpha = f,$$

and

$$\gamma(p, q, p') = f(p) + q.$$

Then $\alpha$ and $\beta$ are split injective quasi-isomorphisms. Using the second case, we can now define

$$\det f = (\det \beta)^{-1}(\det \alpha).$$

Finally, if $f$ and $g$ are two quasi-isomorphisms $P. \to Q.$, which are homotopic, say by

$$f - g = dh + hd, \qquad\qquad h : P. \longrightarrow Q.[1],$$

there is an isomorphism
$$Z^f \xrightarrow{\sim} Z^g$$
$$(p, q, p') \longmapsto (p, q + hp', p')$$
such that the following diagram commutes:

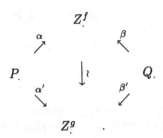

This implies
$$(\det \beta)^{-1} \det \alpha = (\det \beta')^{-1} \det \alpha'.$$

So $\det f = \det g$. The proof of Lemma 1 is now complete (except that we should check that the definitions in the first and second cases are compatible with the general procedure of the third case; this is left to the reader).                                                       □

**1.3**    The construction we have just made is a special case of the following theorem of Grothendieck and Knudsen–Mumford [KM]:

**Theorem 1**    *There is one and, up to canonical isomorphism, only one functor:*

$$\left\{ \begin{array}{c} \textit{finite complexes of locally} \\ \textit{free coherent } \mathcal{O}_X - \textit{modules} \\ \textit{and quasi-isomorphisms} \end{array} \right\} \xrightarrow{\det} \left\{ \begin{array}{c} \textit{graded line bundles} \\ \textit{and} \\ \textit{isomorphisms} \end{array} \right\}$$

*together with isomorphisms*
$$\det S_. \otimes \det Q_. \xrightarrow{\varphi_{\mathcal{E}}} \det E_.$$

*for any short exact sequence $\mathcal{E} : 0 \to S_. \to E_. \to Q_. \to 0$ such that:*

(i)    *they are functorial in $X$, i.e. they commute with base change;*

(ii)   *$\det$ and $\varphi$ are as expected in the following cases:*
  - *if $P_. = 0$, then $\det P_. = 1_X$;*
  - *if $\mathcal{E}$ is $0 \to E_. \xrightarrow{\mathrm{id}} E_. \to 0$ or $0 \to 0 \to E_. \xrightarrow{\mathrm{id}} E_. \to 0$, then $\varphi_{\mathcal{E}} = \mathrm{id}$;*
  - *if $P_. = (0 \to P_0 \to 0)$, then $\det P_. = \det P_0$ as defined before; similarly for $\varphi_{\mathcal{E}}$ of any exact sequence*
    $$\mathcal{E} : \quad 0 \to S_. \to E_. \to Q_. \to 0$$

with $S_.,E_.,Q_.$ concentrated in degree 0;

(iii)   a commutative diagram

$$\mathcal{E}_1 : \quad 0 \longrightarrow P'_. \longrightarrow P_. \longrightarrow P''_. \longrightarrow 0$$
$$\downarrow f' \qquad \downarrow f \qquad \downarrow f''$$
$$\mathcal{E}_2 : \quad 0 \longrightarrow Q'_. \longrightarrow Q_. \longrightarrow Q''_. \longrightarrow 0,$$

where $f', f''$, and $f$ are quasi-isomorphisms, gives rise to a commutative diagram

$$\det P'_. \otimes \det P''_. \overset{\varphi_{\mathcal{E}_1}}{\longrightarrow} \det P_.$$
$$\downarrow \wr \qquad\qquad\qquad \downarrow \wr$$
$$\det Q'_. \otimes \det Q''_. \overset{\varphi_{\mathcal{E}_2}}{\longrightarrow} \det Q_. \ ;$$

(iv)   given a commutative exact square

$$
\begin{array}{ccccccccc}
 & & 0 & & 0 & & 0 & & \\
 & & \downarrow & & \downarrow & & \downarrow & & \\
\mathcal{E}' : & 0 \longrightarrow & S'_. & \longrightarrow & E'_. & \longrightarrow & Q'_. & \longrightarrow & 0 \\
 & & \downarrow & & \downarrow & & \downarrow & & \\
\mathcal{E} : & 0 \longrightarrow & S_. & \longrightarrow & E_. & \longrightarrow & Q_. & \longrightarrow & 0 \\
 & & \downarrow & & \downarrow & & \downarrow & & \\
\mathcal{E}'' : & 0 \longrightarrow & S''_. & \longrightarrow & E''_. & \longrightarrow & Q''_. & \longrightarrow & 0 \\
 & & \downarrow & & \downarrow & & \downarrow & & \\
 & & 0 & & 0 & & 0 & & \\
 & & \mathcal{E}_S & & \mathcal{E}_E & & \mathcal{E}_Q & &
\end{array}
$$

the composite of the isomorphisms

$$(\det S'_.) \otimes (\det S''_.) \otimes (\det Q'_.) \otimes (\det Q''_.)$$
$$\overset{\varphi_{\mathcal{E}_S} \otimes \varphi_{\mathcal{E}_Q}}{\longrightarrow} (\det S_.) \otimes (\det Q_.)$$
$$\overset{\varphi_{\mathcal{E}}}{\longrightarrow} \det E_.$$

is equal to the composite of the isomorphisms

$$(\det S'_.) \otimes (\det S''_.) \otimes (\det Q'_.) \otimes (\det Q''_.)$$
$$\overset{\tau_{(S''_.,Q'_.)}}{\longrightarrow} (\det S'_.) \otimes (\det Q'_.) \otimes (\det S''_.) \otimes (\det Q''_.)$$
$$\overset{\varphi_{\mathcal{E}'} \otimes \varphi_{\mathcal{E}''}}{\longrightarrow} \det E'_. \otimes \det E''_. \overset{\varphi_{\mathcal{E}}}{\longrightarrow} \det E_. .$$

**1.4**   Given any scheme $X$, a complex of $\mathcal{O}_X$-modules is called *perfect* if, locally on $X$, it is quasi-isomorphic to a finite complex of locally free coherent $\mathcal{O}_X$-modules.

**Theorem 2** ([KM])   *There is a unique way (up to canonical isomorphism) of extending the functor* det *and the isomorphisms* $\varphi$ *defined in*

*Theorem 1 to the full subcategory of the derived category of $\mathcal{O}_X$-modules generated by perfect complexes, in such a way that axioms analogous to (i)–(iv) hold.*

If $X$ is regular, noetherian, separated, a coherent $\mathcal{O}_X$-module is a special case of a perfect complex; see the proof of Lemma I.1. If $f : X \to Y$ is a morphism of schemes, proper and of finite Tor dimension, then, for any perfect complex $\mathcal{F}$ of $\mathcal{O}_X$-modules, $Rf_*\mathcal{F}$ is a perfect complex of $\mathcal{O}_Y$-modules (see [GBI], exposé 3, Prop. 4.8). So we can make the following definition:

**Definition 2**    *The determinant of the cohomology is the (graded) line bundle $\lambda(\mathcal{F}) = \det(Rf_*\mathcal{F})$ on $Y$.*

Specializing to the case where $X$ and $Y$ are noetherian and separated, $Y$ is regular, $f : X \to Y$ is a proper map, and $E$ is a coherent $\mathcal{O}_X$-module, we get

$$\lambda(E) = \bigotimes_{q \geq 0}(\det R^q f_* E)^{(-1)^q},$$

where $\det R^q f_* E$ is the determinant of the coherent $\mathcal{O}_Y$-module $R^q f_* E$, as defined in §1.2. If $f$ is flat and $X_y$ denotes the fiber at $y \in Y$ we have

$$(1) \qquad \lambda(E)_y = \bigotimes_{q \geq 0}(\det H^q(X_y, E))^{(-1)^q}.$$

Finally, we note that all of this discussion carries over without difficulty to the analytic category. In particular, if $f : X \to Y$ is a proper flat map of complex manifolds and $E$ a holomorphic vector bundle on $X$, there exists a holomorphic line bundle $\lambda(E)$ on $Y$, called the *determinant of the cohomology*, with natural isomorphisms as (1) above.

## 2. The determinant line bundle: analytic approach

**2.1**    Let $f : X \to Y$ be a smooth proper map between complex varieties, $E$ a holomorphic vector bundle on $X$. We assume furthermore that each fiber $X_y = f^{-1}(y)$, $y \in Y$ is endowed with a Kähler metric, varying smoothly with $y$.

Choose an hermitian metric on $E$. Then we may consider the Laplace operator $\Delta_y^q = \overline{\partial}\overline{\partial}^* + \overline{\partial}^*\overline{\partial}$ acting on $A^{0,q}(X_y, E)$ (see §V.2.2). Hodge theory for the Dolbeault resolution of $E$ on $X_y$ gives us an isomorphism

$$H^q(X_y, E) \xrightarrow{\sim} \ker \Delta_y^q$$

for all $q \geq 0$. In general, the vector spaces $\ker \Delta_y^q$, $y \in Y$, are *not* the fibers of a vector bundle "ker $\Delta^q$" on $Y$. Indeed, their dimension is not always locally constant.

Consider, for example, the relative dimension one case. We then have the Dolbeault complex

$$A^{0,0}(X_y, E) \xrightarrow{\ \bar{\partial}\ } A^{0,1}(X_y, E)$$

so that $\Delta_y^0 = \bar{\partial}^* \bar{\partial}, \Delta_y^1 = \bar{\partial}\bar{\partial}^*, H^0 \simeq \ker \Delta_y^0$ and $H^1 \simeq \ker \Delta_y^1$. Assume $v \in A^{0,0}(X_y, E)$ is a non-zero eigenvector of $\Delta_y^0$ with *non-zero* eigenvalue $\lambda$. Then

$$\Delta_y^1(\bar{\partial}v) = \bar{\partial}\,\bar{\partial}^*\bar{\partial}v = \bar{\partial}\,\lambda v = \lambda\,\bar{\partial}v.$$

Moreover $\bar{\partial}v \neq 0$ because

$$\Delta_y^0 v = \bar{\partial}^*\bar{\partial}v = \lambda v \neq 0.$$

This shows that the non-zero eigenspaces of $\Delta_y^0$ and $\Delta_y^1$ are set in bijection by $\bar{\partial}$. If, in a family, $v_y$ becomes a null-vector for $\Delta_{y_0}^0$, then so does $\bar{\partial}v_y$ for $\Delta_{y_0}^1$. So even if $H_y^0$ and $H_y^1$ may "jump" at $y_0$, $(\Lambda^{\max} H_y^0) \otimes (\Lambda^{\max} H_y^1)^{-1}$ does not.

In the following we will develop these ideas to define analytically a smooth line bundle $\lambda(E)_\infty$ on $Y$, with natural isomorphisms, for all $y \in Y$,

$$\lambda(E)_{\infty,y} \simeq \bigotimes_{q \geq 0} \det H^q(A^{0,\cdot}(X_y, E), \bar{\partial}_{X_y})^{(-1)^q}.$$

Let $b \in \mathbb{R}$, $b > 0$, and let $K_y^{q,b}$ be the subspace of $A^{0,q}(X_y, E)$ generated by the eigenvectors of $\Delta_y^q$ with eigenvalues strictly less than $b$. Denote by $\mathrm{Sp}(\Delta_y^q)$ the spectrum of $\Delta_y^q$.

**Proposition 1**    *Assume there exists $\epsilon > 0$ such that, for all $y \in Y$ and $q \geq 0$,*

$$\mathrm{Sp}(\Delta_y^q) \cap ]b - \epsilon, b + \epsilon[ = \emptyset.$$

*Then there is a smooth vector bundle $K^{q,b}$ on $Y$ with fibers $K_y^{q,b}$.*

*Proof.* Choose a smooth function $\varphi$ in $\mathbb{R}$ such that

$$\varphi(s) = 1 \qquad \text{for} \quad \frac{|s|^2}{2} \leq b - \epsilon,$$

$$\varphi(s) = 0 \qquad \text{for} \quad \frac{|s|^2}{2} \geq b + \epsilon.$$

Let

$$\psi(s) = e^{-s^2}\varphi(s)$$

and

$$\hat{\psi}(\sigma) = \frac{1}{2\pi} \int_{\mathbb{R}} \psi(s) e^{-i\sigma s} ds.$$

Consider the Dirac operators $D_y = \sqrt{2}(\bar\partial + \bar\partial^*)$. For any $\sigma \in \mathbf{R}$, $D_y^2 - i\sigma D_y$ is a generalized Laplacian on $X_y$, acting on

$$V_y = \bigoplus_{q \geq 0} A^{0,q}(X_y, E).$$

So the operator $e^{i\sigma D_y - D_y^2}$ is well defined. It has a smooth kernel which, according to §V.3.6.1, depends smoothly on $y$ and $\sigma$. However, we need more work to prove the following:

**Lemma 2**    *The integral $\int_{\mathbf{R}} \hat\psi(\sigma)e^{i\sigma D_y - D_y^2} d\sigma$ converges and defines an operator $\varphi(D_y)$ with a smooth kernel depending smoothly on $y$.*

We postpone the proof till §2.2 below.

**Lemma 3**    $\varphi(D_y)$ *is the orthogonal projection operator onto*

$$\bigoplus_{q \geq 0} K_y^{q,b}.$$

*Proof.*   Indeed, using $D_y^2 = 2 \bigoplus_{q \geq 0} \Delta_y^q$, we see that

$$V_y = \bigoplus_{q \geq 0} A^{0,q}(X_y, E)$$

has an orthogonal decomposition in eigenspaces of $D_y$: $V_y = \bigoplus_\lambda V_{y,\lambda}$. For $v \in V_{y,\lambda}$, $e^{(i\sigma D_y - D_y^2)}v = e^{i\sigma\lambda - \lambda^2}v$ (say, by uniqueness of a solution to a first order differential equation), so

$$\varphi(D_y)v = \left( \int_{\mathbf{R}} \hat\psi(\sigma)e^{i\sigma\lambda} d\sigma \right) e^{-\lambda^2} v = \varphi(\lambda)v.$$

Hence

$$\varphi(D_y)v = v \qquad \text{if } \frac{\lambda^2}{2} \leq b - \epsilon$$

and

$$\varphi(D_y)v = 0 \qquad \text{if } \frac{\lambda^2}{2} \geq b + \epsilon.$$

But the $\frac{\lambda^2}{2}$ are the eigenvalues of $\bigoplus_{q \geq 0} \Delta_y^q$. Lemma 3 now follows from the assumption that there is no eigenvalue of $\Delta_y^q$ in $]b - \epsilon, b + \epsilon[$.

**Lemma 4**    dim $K_y^{q,b}$ *is locally constant in $y$.*

*Proof.*   Let $P_y^{q,b} : A^{0,q}(X_y, E) \to K_y^{q,b}$ be the orthogonal projection operator. According to Lemmas 2 and 3, it depends smoothly on $y$. Now dim $K_y^{q,b} = \mathrm{tr} P_y^{q,b}$ is an integer and depends smoothly on $y$. The conclusion follows.    □

We can now complete the proof of Proposition 1. Let $(\phi^j)$ be a basis of $K_{y_0}^{q,b}$. We extend the $\phi^j$ to sections $\phi^j(y)$ of $A^{0,q}(X_y, E)$, smooth with respect to $y$. Define $\psi^j(y) = P_y^{q,b}\phi^j(y)$. The $\psi^j(y)$ are sections of $\coprod_{y \in Y} K_y^{q,b}$, depending smoothly on $y$. As $(\psi^j(y_0))_j$ are linearly independent, so are $(\psi^j(y))_j$ for $y$ in a neighborhood $U$ of $y_0$. Thanks to Lemma 4, they actually give a basis of $K_y^{q,b}$ in this neighborhood. It is clear that changes in our choices would lead to sections $(\theta^j(y))_j$ related to the previous ones by linear relations depending smoothly on $y$. But this says exactly that the collection of isomorphisms

$$\coprod_{y \in U} K_y^{q,b} \xrightarrow{\sim} U \times \mathbb{C}^{\mathrm{rank}\, K_{y_0}^{q,b}}$$

glue together to define a smooth vector bundle $K^{q,b}$ on $Y$, with fibers the $K_y^{q,b}$'s.  $\square$

**2.2**   We shall now prove Lemma 2.

If $\alpha$ is any natural integer, we denote by $\|.\|_\alpha$ any $\alpha$-norm, i.e. a sup-norm on $X_y$ (resp. $X \times_Y X$) involving $\alpha$ derivatives (resp. along the fibers of the projection to $Y$). If $K$ is an operator with smooth kernel $k(x, x')$ and $\alpha$ a natural integer, we shall write $\|K\|_\alpha$ for $\|k(x, x')\|_\alpha$ where $\|.\|_\alpha$ is any $\alpha$-norm, i.e. a sup-norm on $X \times_Y X$ involving $\alpha$ derivatives along the fibers. Furthermore, $\|K\|_{L^p}$ stands for the $L^p$ norm of $k(x, x')$ (it is *not* the $L^p$-norm of the operator $K$).

**2.2.1**   We first prove three lemmas which will be used later. Throughout this section, $y$ is assigned to vary in an open set $U$ with compact closure in $Y$. Let $n = \dim_{\mathbb{C}} X_y$ and $H_{y,\sigma} = \Delta_y - i\sigma D_y$ where we write $\Delta_y$ instead of $\Delta_y^q$ until §2.3.

**Lemma 5**   *For any $\alpha$-norm and any $(\alpha + n + 2)$-norm on $X_y$, depending continuously on $y$, there exists a constant $C$ such that, for every $\varphi \in \Gamma(f^{-1}(U), E)$, $u \geq 0$, $\sigma \in \mathbb{R}$, and $y \in U$, the following holds:*

$$\|e^{-uH_{y,\sigma}}\varphi\|_\alpha \leq C\|\varphi\|_{\alpha+n+2}.$$

*Proof.*   By Sobolev's lemma and Garding's inequality, applied to the elliptic operator $\Delta_y$ acting on $X_y$, it is sufficient to prove under the same conditions that there exists $C$ such that, if $2k \leq \alpha + n + 2$ and if $\Delta_y^k$ is the composite of $k$ copies of $\Delta_y$,

$$\|\Delta_y^k e^{-uH_{y,\sigma}}\varphi\|_{L^2} \leq C\|\varphi\|_{\alpha+n+2}.$$

In fact, $\|e^{-uH_{y,\sigma}}\varphi\|_\alpha \leq C\|e^{-uH_{y,\sigma}}\varphi\|_{H^{\alpha+n+1}}$ by Sobolev's lemma (see

Lemma 1.3.4 of [Gi]; an inspection of the proof shows that $C$ may be taken independent of $y$). And

$$\|e^{-uH_{y,\sigma}}\varphi\|_{H^{\alpha+n+1}} \leq C \sum_{2k=0}^{\alpha+n+2} \|\Delta_y^k e^{-uH_{y,\sigma}}\varphi\|_{L^2}$$

by Garding's inequality (see Lemma 1.3.5 of [Gi]; same remark as above for $C$).

But, for all $u \geq 0$, $\sigma \in \mathbb{R}$, and $y \in U$, the $L^2$-norm of the operator $e^{-uH_{y,\sigma}}$ is less or equal to one, so we get

$$\|\Delta_y^k e^{-uH_{y,\sigma}}\varphi\|_{L^2} = \|e^{-uH_{y,\sigma}}\Delta_y^k\varphi\|_{L^2} \leq \|\Delta_y^k\varphi\|_{L^2} \leq C\|\varphi\|_{\alpha+n+2}.$$

The conclusion follows.                                                  □

**Lemma 6**    *For any family of operators with smooth kernel $K_y$, differential operators $D_y$ of order $\leq d$, any $\alpha$-norm and any $\alpha+d+n+2$-norm on $X_y \times X_y$ (all these depending continuously on $y$), there exists a constant $C$ such that, for any $y \in U$, any $u \geq 0$, and any $\sigma \in \mathbb{R}$,*

$$\|e^{-uH_{y,\sigma}}D_yK_y\|_\alpha \leq C\|K_y\|_{\alpha+d+n+2}$$

*and*

$$\|K_yD_ye^{-uH_{y,\sigma}}\|_\alpha \leq C\|K_y\|_{\alpha+d+n+2}.$$

*Proof.* The second inequality is reduced to the first one by taking adjoints. By Lemma 5

$$\|e^{-uH_{y,\sigma}}D_yK_y\|_\alpha \leq C\|D_yK_y\|_{\alpha+n+2}$$

and obviously $\|D_yK_y\|_{\alpha+n+2} \leq C\|K_y\|_{\alpha+d+n+2}$. Note that $C$ does not depend on the $K_y$'s.                                         □

**Lemma 7**    *For any $\alpha$-norm on $X_y \times X_y$ depending continuously on $y$, and for any constants $A > B > 0$, there exists a constant $C$ independent of $y$ such that, if $u \in [B, A]$, for any $\sigma \in \mathbb{R}$,*

$$\|e^{-uH_{y,\sigma}}\|_\alpha \leq C.$$

*Proof.* Applying Sobolev's lemma and Garding's inequality (cf. 1.3.4 and 1.3.5 in [Gi]) to the elliptic operator $\Delta_y \otimes 1 + 1 \otimes \Delta_y$ on $X_y \times X_y$ and using the fact that $\Delta_y$ is self-adjoint, and commutes with $e^{-H_{y,\sigma}}$, one is reduced to proving, for all $k \geq 0$,

$$\|\Delta_y^k e^{-H_{y,\sigma}}\|_{L^2} \leq C$$

where $C$ depends on $k$ but not on $y$, $u$ or $\sigma$.

But in some Hilbert space completion of $V_y = \bigoplus_{q \geq 0} A^{0,q}(X_y, E)$, there exists a *unitary* operator $e^{i\sigma D_v}$, commuting with $\Delta_y$ and such that

$$e^{-H_{v,\sigma}} = e^{i\sigma D_v} e^{-\Delta_v}.$$

Therefore the quantity of interest is bounded by $\|\Delta_y^k \, e^{-u\Delta_v}\|_{L^2}$, which is less than $C\|e^{-u\Delta_v}\|_{2k}$, where $\|\cdot\|_{2k}$ is any $2k$-norm on $X \times_U X$. But the kernel of $e^{-u\Delta_v}$ depends smoothly on $(u, y)$ in $]0, +\infty[\times U$. So the conclusion follows.  □

**2.2.2**  Let

$$I = \int_{\mathbb{R}} \hat{\psi}(\sigma) e^{-H_{v,\sigma}} d\sigma.$$

For a fixed $y$, the integral $I$ converges and defines a smooth kernel, hence an operator $\varphi(D_y)$. To see that, it is sufficient to prove that for any $\alpha$-norm on $X_y \times X_y$ there is a constant $C$, depending on $\alpha$ but not on $\sigma$ such that

$$\|e^{-H_{v,\sigma}}\|_\alpha \leq C.$$

But this follows from Lemma 7.

We will now prove that this kernel of $\varphi(D_y)$ depends smoothly on $y$. If $\frac{\partial}{\partial y}$ is any local derivative on $Y$, we get, from formula (16) of §V.3.6.3, that the derivative of (the kernel of) $e^{-H_{v,\sigma}}$ is given by the integral

$$\frac{\partial}{\partial y} e^{-H_{v,\sigma}} = -\int_0^1 e^{-(1-s)H_{v,\sigma}} \frac{\partial}{\partial y}(H_{y,\sigma}) e^{-sH_{v,\sigma}} ds.$$

By iterating this formula (or by using Duhamel's formula (15) from Chapter V) we obtain that, given $r$ derivatives, the kernel of

$$\frac{\partial}{\partial y_1} \cdots \frac{\partial}{\partial y_r} e^{-H_{v,\sigma}}$$

is a finite alternated sum of integrals

$$(2) \qquad \sigma^k \int_{\Delta_1^k} e^{-(1-t_k)H_{v,\sigma}} A_k e^{-(t_k - t_{k-1})H_{v,\sigma}} \cdots A_1 e^{-t_1 H_{v,\sigma}} dt_1 \cdots dt_k.$$

The sum is taken over all sequences $n_1, \cdots, n_k$ of positive integers such that $\sum_{j \geq 0} n_j = r$, and $A_j$ is the composite of $n_j$ derivatives of $-iD_y$. In particular $A_j$ is a differential operator independent of $\sigma$. We shall prove that the integrand in (2) is bounded in sup-norm independently of $\sigma$. Therefore

$$(3) \qquad \left\| \frac{\partial}{\partial y_1} \cdots \frac{\partial}{\partial y_r} e^{-H_{v,\sigma}} \right\|_0 \leq C(|\sigma|^r + 1)$$

hence, since $\hat{\psi}(\sigma)$ is in the Schwartz class, the kernel of

$$\varphi(D_y) = \int \hat{\psi}(\sigma) \, e^{-H_{v,\sigma}} d\sigma$$

depends smoothly on $y$.

To prove that the integrand in (2) is bounded independently of $\sigma$ we choose an index $j$ such that

$$t_j - t_{j-1} \geq \frac{1}{k} \geq \frac{1}{r}.$$

If we apply Lemma 6 several times, with $D_y$ equal to $A_1, \cdots, A_k$, we get that

(4)
$$\begin{aligned} \|e^{-(1-t_k)H_{y,\sigma}} A_k e^{-(t_k-t_{k-1})H_{y,\sigma}} \cdots A_1 e^{-t_1 H_{y,\sigma}}\|_0 \\ \leq C \|e^{-(t_j-t_{j-1})H_{y,\sigma}}\|_{k(n+3)}. \end{aligned}$$

We then apply Lemma 7 to see that that the right-hand side of (4) is independent of $\sigma$. The proof of Lemma 2 is now complete.    □

**Remark**    Our proofs of Proposition 1 and Lemma 2 are inspired by a preliminary version of [BGV], Chapter 9. The reader should consult *loc. cit.* for a different and more detailed treatment.

## 2.3

**Lemma 8**    *For any $b > 0$ and $\epsilon > 0$ the subset*

$$U^{b,\epsilon} = \{y \in Y \mid \forall q \geq 0 \quad \mathrm{Sp}(\Delta_y^q) \cap [b - \epsilon, b + \epsilon] = \emptyset\}$$

*is open in $Y$.*

*Proof.* Choose $\varphi$ a smooth function on $\mathbb{R}$ with values in $[0, 1]$, with compact support, and taking the value 1 exactly on $[b - \epsilon, b + \epsilon]$. As in Lemma 2, we define

$$\varphi(D_y) = \int_{\mathbb{R}} \hat{\psi}(\sigma) e^{i\sigma D_y - D_y^2} d\sigma \qquad \text{with } \psi(s) = e^{s^2} \varphi(s).$$

This operator depends smoothly on $y$ and $\mathrm{Sp}(\Delta_y^q) \cap [b - \epsilon, b + \epsilon] = \emptyset$ for at least one $q$ if and only if there exists $v$ such that $\varphi(D_y)(v) = v$, i.e. if and only if $\|\varphi(D_y) - I\| = 0$ (operator norm). But this is a closed condition (note that $\varphi(D_y)$ is bounded) so the conclusion follows.    □

Using Proposition 1 and Lemma 8 we now cover $Y$ by open sets $U^{b,\epsilon}$ over which all $K^{q,b}, q \geq 0$, are smooth vector bundles. On $U^{b,\epsilon}$ we define a smooth determinant line bundle by the formula

$$\lambda(E)_\infty^{U^{b,\epsilon}} := \bigotimes_{q \geq 0} (\det K^{q,b})^{(-1)^q}.$$

**Proposition 2**    *The $\lambda(E)_\infty^{U^{b,\epsilon}}$ glue together to form a $C^\infty$ line bundle $\lambda(E)_\infty$ on $Y$ with fibers $\lambda(E)_y = \bigotimes_{q \geq 0} (\det H^q(X_y, E))^{(-1)^q}$. This line* bundle $\lambda(E)_\infty$ is the $C^\infty$ line bundle underlying the holomorphic line bundle $\lambda(E)$ defined in §1.4.

*Proof.* As $\bar\partial\Delta = \Delta\bar\partial$, the map $\bar\partial$ maps $K^{q,b}$ into $K^{q+1,b}$ and we get a complex $(K^{\cdot,b}, \bar\partial)$. If $b < b'$ there is an exact sequence of complexes

$$0 \longrightarrow (K^{\cdot,b}, \bar\partial) \longrightarrow (K^{\cdot,b'}, \bar\partial) \longrightarrow (Q^\cdot, \bar\partial) \longrightarrow 0$$

where $(Q^\cdot, \bar\partial)$ is acyclic since

$$H^q(Q^\cdot, \bar\partial) = \ker \Delta^q\big|_{Q^\cdot} = 0.$$

Therefore, on $U^{b,\epsilon} \cap U^{b',\epsilon'}$, we have a canonical isomorphism

$$\bigotimes_{q\geq 0} (\det K^{q,b})^{(-1)^q} \xrightarrow{\ \sim\ } \bigotimes_{q\geq 0} (\det K^{q,b'})^{(-1)^q}$$

with natural compatibilities on triple intersections. This proves the first assertion.

To compare $\lambda(E)$ and $\lambda(E)_\infty$, let $D_X^\cdot$ be the *total* Dolbeault complex, i.e. the complex of $\mathcal{O}_Y$-sheaves associated to the presheaf

$$U \longmapsto (A^{0,\cdot}(f^{-1}(U), E), \bar\partial_U).$$

By the Dolbeault isomorphism we have

$$\lambda(E) = \det(Rf_*E) \simeq \det(D_X^\cdot).$$

On the other hand, let $D_f^\cdot$ be the *relative* Dolbeault complex, i.e. the complex of $\mathcal{O}_Y^\infty$-sheaves obtained from the presheaf of smooth sections

$$U \longmapsto \Gamma(f^{-1}(U), \Lambda^q(Tf^{*(0,1)}) \otimes E, \bar\partial),$$

where $Tf^{*(0,1)}$ denotes the relative differentials of type $(0,1)$, and $\bar\partial$ the Cauchy–Riemann operator along the fibers of $f$. This complex $(D_f^\cdot, \bar\partial)$ is a perfect complex since on $U^{b,\epsilon}$ the inclusion $(K^{\cdot,b}, \bar\partial) \to (D_f^\cdot, \bar\partial)$ is a quasi-isomorphism. We have $\lambda(E)_\infty = \det(D_f^\cdot)$.

There is a natural map of Dolbeault complexes:

$$D_X^\cdot \otimes_{\mathcal{O}_Y} \mathcal{O}_Y^\infty \longrightarrow D_f^\cdot.$$

The complex $D_X^\cdot$ is perfect (since $Rf_*E$ is perfect). This implies [BGS1] that, locally on $Y$, we can find a bounded complex $P^\cdot$ of finite-dimensional vector bundles and a quasi-isomorphism $P^\cdot \longrightarrow D_X^\cdot$. Since $P^\cdot$ is a complex of vector bundles, there is a morphism of complexes $p : P^\cdot \longrightarrow K^{\cdot,b}$ making the diagram

$$
\begin{array}{ccc}
P^\cdot & \longrightarrow & D_X^\cdot \\[1em]
{\scriptstyle p}\downarrow & & \downarrow \\[1em]
K^{\cdot,b} & \longrightarrow & D_f^\cdot
\end{array}
$$

commutative. Now, the induced map

$$p^\infty : P^{\cdot} \otimes_{\mathcal{O}_Y} \mathcal{O}_Y^\infty \longrightarrow K^{\cdot, b}$$

is a quasi-isomorphism because at any point $y$ it induces the identity map

$$H^q(X_y, E) \xrightarrow{\text{id}} H^q(X_y, E)$$

on cohomology. So we have an isomorphism

$$\det(P^{\cdot} \otimes_{\mathcal{O}_Y} \mathcal{O}_Y^\infty) \xrightarrow{\det p^\infty} \det(K^{\cdot, b}).$$

Now $\det(P^{\cdot} \otimes_{\mathcal{O}_Y} \mathcal{O}_Y^\infty) = (\det P^{\cdot}) \otimes_{\mathcal{O}_Y} \mathcal{O}_Y^\infty$ since, by a theorem of Malgrange ([Mg], Thm. 4), $\mathcal{O}_Y^\infty$ is flat over $\mathcal{O}_Y$. Furthermore

$$\det P^{\cdot} \xrightarrow{\sim} \det(D_X^{\cdot}) \quad \text{and} \quad \det K^{\cdot, b} \xrightarrow{\sim} \det(D_f^{\cdot}).$$

This gives a local isomorphism $\det(D_X^{\cdot}) \otimes_{\mathcal{O}_Y} \mathcal{O}_Y^\infty \xrightarrow{\sim} \det(D_f^{\cdot})$, and these local isomorphisms glue together. Finally the map

$$\lambda(E) \otimes_{\mathcal{O}_Y} \mathcal{O}_Y^\infty \longrightarrow \lambda(E)_\infty$$

is an isomorphism, as was to be proved; for more details, see [BGS1].

$$\square$$

## 3. Quillen's metric

**3.1**    Let $f : X \to Y$ be a smooth proper map of complex manifolds. Let $E$ be a holomorphic vector bundle on $X$. Suppose moreover that there is an hermitian metric $h$ on $E$ and an hermitian metric $h_f$ on $Tf := \ker(TX \to f^*TY)$, i.e. a metric on each fiber $X_y = f^{-1}(y)$, varying smoothly with $y$. Furthermore we assume that the restriction of $h_f$ to $X_y$ is Kähler for every $y \in Y$.

With these data we define as follows a natural metric, called the $L^2$-*metric* on each fiber of the determinant line bundle $\lambda(E)$. Recall that, for every $y \in Y$,

$$\lambda(E)_y = \bigotimes_{q \geq 0} (\det H^q(X_y, E))^{(-1)^q}.$$

Now, the Hodge theorem gives an isomorphism

$$H^q(X_y, E) \xrightarrow{\sim} \ker(\Delta_y^q)$$

where

$$\Delta_y^q = \overline{\partial}\, \overline{\partial}^* + \overline{\partial}^* \overline{\partial}$$

is the Laplace operator acting on $A^{0,q}(X_y, E)$. Furthermore $A^{0,q}(X_y, E)$

has an $L^2$-scalar product given, as in §V.2.2, by the formula

(5)
$$\langle s,t\rangle_{L^2} = \int_{X_y} \langle s(x), t(x)\rangle \frac{\omega_0^n}{n!},$$

where $n = \dim_{\mathbb{C}} X_y$, and $\omega_0$ is the normalized Kähler form, given in any local chart $(z_\alpha)$ on $X_y$ by

$$\omega_0 = \frac{i}{2\pi} \sum_{\alpha,\beta} h_f\left(\frac{\partial}{\partial z_\alpha}, \frac{\partial}{\partial z_\beta}\right) dz_\alpha \overline{dz_\beta}.$$

So we get an $L^2$ metric on $H^q(X_y, E)$, hence on $\lambda(E)_y$. But in general, this metric is *not* smooth in $y$ (not even continuous).

To see what happens, let us consider the case of relative dimension one. Recall that

$$\bar{\partial} : A^{0,0}(X_y, E) \to A^{0,1}(X_y, E)$$

induces isomorphisms

(6)
$$\ker(\Delta_y^0 - \lambda \text{ id}) \xrightarrow{\sim} \ker(\Delta_y^1 - \lambda \text{ id})$$

for any $\lambda > 0$. Now, when $v \in \ker(\Delta_y^0 - \lambda \text{ id})$, one computes

$$\langle \bar{\partial}v, \bar{\partial}v\rangle_{L^2} = \langle \bar{\partial}^*\bar{\partial}v, v\rangle_{L^2} = \langle \Delta_y^0 v, v\rangle_{L^2} = \lambda\langle v, v\rangle_{L^2}.$$

So the $L^2$-norm of the isomorphism (6) is $\sqrt{\lambda}$. For any $b > 0$, consider the isomorphism

(7)
$$\begin{aligned} i_b : \lambda(E)_y &= (\det \ker \Delta_y^0) \otimes (\det \ker \Delta_y^1)^{-1} \\ &\xrightarrow{\sim} (\det K_y^{0,b}) \otimes (\det K_y^{1,b})^{(-1)}. \end{aligned}$$

For the $L^2$-metric induced from $A^{0,q}(X_y, E)$, the computation made above gives:

(8)
$$\|i_b\|_{L^2} = \prod_{0<\lambda\leq b}\left(\frac{1}{\sqrt{\lambda}}\right).$$

So $i_b$ is not an isometry in general.

If $b$ is not in $\text{Sp}(\Delta_{y_0}^q)$, we know that in a neighborhood of $y_0$ the vector spaces $K_y^{q,b}$ are fibers of a smooth vector bundle; see the discussion before Proposition 2. So the natural $L^2$-metric on $\det K_y^{\cdot,b}$ is smooth with respect to $y$. Therefore the failure for the $L^2$-metric on $\lambda(E)_y$ to be smooth (or continuous) with respect to $y$, is given exactly by the behavior of

$$\prod_{0<\lambda\leq b} \lambda =: \det_{]0,b]} \Delta,$$

with $\Delta = \Delta_y^1$.

**3.2**    The previous discussion leads us to introduce new metrics on

both sides of (7). On $\lambda(E)_y$ we define
$$\|\cdot\|_Q^2 = \|\cdot\|_{L^2}^2 \times (\det{}_{>0}\Delta)^{-1}$$
and on the right-hand side of (7) we consider $\|\cdot\|_Q^2 = \|\cdot\|_{L^2}^2 \times (\det{}_{>b}\Delta)^{-1}$, where $\det{}_{>b}$ is the determinant of the restriction to the orthogonal complement of $K_y^{0,b}$. For these new metrics, we get from (8) that
$$\|i_b\|_Q^2 = (\det{}_{]0,b]}\Delta)^{-1}(\det{}_{>b}\Delta)^{-1}(\det{}_{>0}\Delta).$$
By Lemma V.1, this gives $\|i_b\|_Q^2 = 1$.

In arbitrary dimension, following Quillen [Q2], we make the:

**Definition 3**     *The Quillen metric on the determinant of the cohomology is*
$$h_Q = h_{L^2} \times \prod_{q\geq 0}(\det{}'\Delta_y^q)^{(-1)^q q}.$$

In other words,
$$h_Q = h_{L^2} \times \exp(T(E))$$
where $T(E)$ is (twice the logarithm of) the Ray–Singer analytic torsion [RS]:
$$T(E) = \sum_{q\geq 0}(-1)^{q+1}q\zeta_q'(0),$$
with $\zeta_q(s) = \mathrm{tr}'(\Delta_y^q)^{-s}$; see §V.4.3.

This section and the next chapter will be devoted to a proof of the following results:

**Theorem 3**     *The metric $h_Q$ is smooth on $\lambda(E)$.*

**Theorem 4**     *Assume that $f$ is locally Kähler, i.e. for every $y \in Y$ there is a neighborhood $U$ of $y$ such that $f^{-1}(U)$ has a Kähler metric. Then*
$$c_1(\lambda(E)_Q) = f_*(ch(E,h)Td(Tf,h_f))^{(2)}$$
*where $Td(Tf,h_f)$ is the Todd form of $(Tf,h_f)$ and $(.)^{(2)}$ denotes the component of degree 2.*

**3.3**     We first prove a corollary of Theorem 4.

**Corollary 1**     *Let $\overline{\mathcal{E}} : 0 \to (S,h') \to (E,h) \to (Q,h'') \to 0$ be an exact sequence of hermitian vector bundles on $X$. Then the isomorphism $\varphi_{\mathcal{E}} : \lambda(S) \otimes \lambda(Q) \to \lambda(E)$ is such that*
$$\log\|\varphi_{\mathcal{E}}\|_Q^2 = f_*(\widetilde{ch}(\overline{\mathcal{E}})Td(f,h_f))^{(0)},$$
*where $\widetilde{ch}(\overline{\mathcal{E}})$ is the Bott–Chern secondary characteristic class defined in Theorem IV.2.*

*Proof.* Consider the diagram:

$$
\begin{array}{ccc}
X \times \mathbf{P}^1 & \longrightarrow & X \\
\Big\downarrow{\tilde{f}} & & \Big\downarrow{f} \\
Y \times \mathbf{P}^1 & \longrightarrow & Y.
\end{array}
$$

On $X \times \mathbf{P}^1$ the bundle

$$
\tilde{E} = (S(1) \oplus E)/S
$$

sits inside an exact sequence

$$
0 \to S(1) \to \tilde{E} \to Q \to 0;
$$

see the proof of Theorem IV.2. Let us choose a metric $\tilde{h}$ on $\tilde{E}$ such that

$$
(\tilde{E}, \tilde{h})_{X \times \{0\}} = (E, h),
$$

$$
(\tilde{E}, \tilde{h})_{X \times \{\infty\}} = (S, h') \overset{\perp}{\oplus} (Q, h'').
$$

Then, by definition of $\tilde{\mathrm{ch}}(\overline{\mathcal{E}})$ and Theorem 4 for $\tilde{f}$, we get

$$
\left( \int_{X/Y} \tilde{\mathrm{ch}}(\overline{\mathcal{E}}) Td(f) \right)^{(0)} = - \left( \int_{X/Y} \int_{\mathbf{P}^1} \mathrm{ch}(\tilde{E}, \tilde{h}) Td(f) \log |z|^2 \right)^{(0)}
$$

$$
= - \int_{\mathbf{P}^1} \left( \int_{X \times \mathbf{P}^1 / Y \times \mathbf{P}^1} \mathrm{ch}(\tilde{E}, \tilde{h}) Td(\tilde{f}) \right)^{(2)} \log |z|^2
$$

$$
= - \int_{\mathbf{P}^1} c_1(\lambda(\tilde{E})_Q) \log |z|^2.
$$

The exact sequence $0 \to S(1) \to \tilde{E} \to Q \to 0$ induces an isomorphism

$$
\tilde{\varphi} : \lambda(S(1)) \otimes \lambda(Q) \overset{\sim}{\longrightarrow} \lambda(\tilde{E}).
$$

At $0$ it coincides with the isomorphism $\varphi_{\mathcal{E}}$ and at $\infty$ this is an isometry. It follows that

$$
c_1(\lambda(\tilde{E})_Q) = c_1(\lambda(S(1))_Q \otimes \lambda(Q)_Q) - dd^c(\log \|\tilde{\varphi}\|_Q^2.
$$

Since $\lambda(S(1))_Q = \lambda(S)_Q(1)$ we get, as in the proof of Theorem IV.2,

$$
\int_{\mathbf{P}^1} c_1(\lambda(S(1))_Q \otimes \lambda(Q)_Q) \log |z|^2 = 0,
$$

and therefore

$$
\left( \int_{X/Y} \tilde{\mathrm{ch}}(\overline{\mathcal{E}}) Td(f) \right)^{(0)} = - \int_{\mathbf{P}^1} c_1(\lambda(\tilde{E})_Q) \log |z|^2
$$

$$
= \int_{\mathbf{P}^1} dd^c(\log \|\tilde{\varphi}\|_Q^2) \log |z|^2
$$

$$
= i_0^* \log \|\tilde{\varphi}\|_Q^2 - i_\infty^* \log \|\tilde{\varphi}\|_Q^2
$$

$$
= \log \|\varphi_{\mathcal{E}}\|_Q^2.
$$

$\square$

**3.4**    A similar formula holds for variation of the metric on the fibers. Let $h_f$ and $h'_f$ be two metrics on $Tf$. Denote by $\widetilde{Td}(f)$ the secondary characteristic class on $X$ such that

$$dd^c\widetilde{Td}(f) = Td(Tf, h_f) - Td(Tf, h'_f),$$

defined in §IV.3.3; see also [BGS1] and [GS3]. Let $h_Q$ (resp. $h'_Q$) be the Quillen metric on $\lambda(E)$ defined using $h_f$ (resp. $h'_f$) on $Tf$ and a fixed metric on $E$.

**Theorem 5**    *Under the above hypotheses*

$$\log(h'_Q/h_Q) = f_*(\text{ch}(E, h)\widetilde{Td}(f))^{(0)}$$

We shall not prove this result; see [BGS1].

**3.4**    We now prove Theorem 3. Since the statement is local on $Y$ we may assume, without loss of generality, that, for some $b > 0$ and for any $y \in Y$ and $q \geq 0$, $b$ is not an eigenvalue of $\Delta_y^q$.

Let

$$\lambda(E)_\infty^b := \bigotimes_{q \geq 0}(\det K^{q,b})^{(-1)^q}.$$

We define a Quillen metric on this line bundle by multiplying the $L^2$-metric by

$$\prod_{q \geq 0}(\det_{>b}\Delta_y^q)^{q(-1)^q}.$$

Let $i_b : \lambda(E)_Q \longrightarrow \lambda(E)_{\infty,Q}^b$ be the natural isomorphism. We will be done if we prove that $i_b$ is an isometry and, for all $q \geq 0$,  $\det_{>b}\Delta_y^q$ depends smoothly on $y$. Indeed, the $L^2$-metric on each $K_y^{q,b}$ is already smooth in $y$.

To prove that $i_b$ is an isometry, we use the Hodge decomposition:

$$A^{0,q} = \ker(\Delta^q) \oplus B^q \oplus C^q$$

where $B^q = \text{im}(\overline{\partial})$ and $C^q = \text{im}(\overline{\partial}^*)$. Notice that $\overline{\partial} : C^q \xrightarrow{\sim} B^{q+1}$ is an isomorphism respecting $\Delta$-eigenspaces. Hence we may define

$$\zeta_{C^q}(s) = \zeta_{B^{q+1}}(s) := \sum_{\substack{\lambda \text{ eigenvalue} \\ \text{of } \Delta \text{ in } C^q}} \lambda^{-s}.$$

Then $\zeta_q(s) = \zeta_{C^q}(s) + \zeta_{C^{q-1}}(s)$. This shows inductively on $q$, starting from $\zeta_0(s) = \zeta_{C^0}(s)$, that the series defining $\zeta_{C^q}(s)$ is convergent for $\text{Re } s > n$, has a meromorphic continuation to the whole complex plane, and no pole at 0.

This implies also

(9)
$$\sum_{q\geq 0}(-1)^{q+1}q\zeta_q(s) = \sum_{q\geq 0}(-1)^q\zeta_{C^q}(s).$$

Defining $\zeta_{C^q,>b}(s)$ in the obvious way, we obtain similarly:

(10)
$$\sum_{q\geq 0}(-1)^{q+1}q\zeta_{q,>b}(s) = \sum_{q\geq 0}(-1)^q\zeta_{C^q,>b}(s).$$

Write $C^{q,b}$ for $C^q \cap K^{q,b}$ and $B^{q,b}$ for $B^q \cap K^{q,b}$. The exact sequence of complexes (with boundary $\bar\partial$)

$$0 \to \ker \Delta^\cdot \to K^{\cdot,b} \to C^{\cdot,b} \oplus B^{\cdot,b} \to 0$$

shows that $i_b$ is the composition

(11)
$$\det(\ker\ \Delta^\cdot) \xrightarrow{\ \mathrm{id}\otimes j_b\ } \det(\ker \Delta^\cdot) \otimes \det(C^{\cdot,b} \oplus B^{\cdot,b})$$

$$i_b \searrow \qquad\qquad\qquad \Big\downarrow \wr$$

$$\det(K^{\cdot,b})$$

where $j_b$ is the natural isomorphism:

$$1 \xrightarrow{\ \sim\ } \det(C^{\cdot,b} \oplus B^{\cdot,b})$$

coming from the acyclicity of $C^{\cdot,b} \oplus B^{\cdot,b}$.

The vertical isomorphism in (11) is an isometry for the $L^2$-metric. The isomorphism $j_b$ decomposes as

$$1 \xrightarrow{\sim} \det(C^{\cdot,b}) \otimes \det(C^{\cdot,b})^{-1} \xrightarrow{\ \mathrm{id}\otimes(\det\ \bar\partial)^{-1}\ } \det(C^{\cdot,b}) \otimes \det(B^{\cdot+1,b})^{-1}$$
$$\xrightarrow{\sim} \det(C^{\cdot,b}) \otimes \det(B^{\cdot,b}) \xrightarrow{\sim} \det(C^{\cdot,b} \oplus B^{\cdot,b}).$$

Here $(\det\ \bar\partial)^{-1}$ is the inverse of $\det\bar\partial : \det C^{q,b} \xrightarrow{\sim} \det B^{q+1,b}$, whose $L^2$-norm, computed at the beginning of the section, equals $\sqrt{\det_{C^{q,b}} \Delta^q}$. So we get

$$\|i_b\|^2_{L^2} = \|j_b\|^2_{L^2} = \prod_{q\geq 0}(\det_{C^{q,b}} \Delta^q)^{(-1)^{q+1}}.$$

This gives

$$\|i_b\|^2_Q =$$
$$\prod_{q\geq 0}(\det_{C^{q,b}} \Delta^q)^{(-1)^{q+1}} \prod_{q\geq 0}(\det_{>b} \Delta^q)^{q(-1)^q} \prod_{q\geq 0}(\det_{>0} \Delta^q)^{q(-1)^{q+1}}.$$

Now from (9) we get

$$\prod_{q\geq 0}(\det_{>0} \Delta^q)^{q(-1)^q} = \prod_{q\geq 0}(\det_{C^q} \Delta^q)^{(-1)^{q+1}}$$

and from (10)

$$\prod_{q\geq 0} (\det{}_{>b} \Delta^q)^{q(-1)^q} = \prod_{q\geq 0} (\det{}_{C^q, >b} \Delta^q)^{(-1)^{q+1}}.$$

Finally

$$\|i_b\|_Q^2 = \frac{\prod_{q\geq 0} (\det_{C^{q,b}} \Delta^q)^{(-1)^{q+1}} \times \prod_{q\geq 0} (\det_{C^q, >b} \Delta^q)^{(-1)^{q+1}}}{\prod_{q\geq 0} (\det_{C^q} \Delta^q)^{(-1)^{q+1}}},$$

i.e. $\|i_b\|_Q = 1$ by Lemma V.1.

**3.6**    Now we prove that $\det_{>b} \Delta_y^q$ is a smooth function of $y$.

**3.6.1**    Let $P_y$ be the orthogonal projection of $A^{0,q}(X_y, E)$ onto $K_y^{q,b}$ and $Q_y = 1 - P_y$. By Lemmas 2 and 3, $P_y$ is an operator with smooth kernel depending smoothly on $y$. So $Q_y$ is smooth in $y$, and $Q_y e^{-t\Delta_y}$ is an operator with smooth kernel for any $t > 0$, depending smoothly on $t$.

Let $\theta_{b,y}(t) = \mathrm{tr}(Q_y e^{-t\Delta_y})$. For technical reasons (see Proposition 1) we assume that $y$ is in an open subset set with compact closure in $Y$ over which $\mathrm{Sp}(\Delta_y) \cap [b - \epsilon, b + \epsilon] = \emptyset$ for a certain $\epsilon > 0$.

For a fixed $y$, $\theta_{b,y}$ satisfies the assumptions $\Theta 1$ and $\Theta 2$ of §V.1. Property $\Theta 1$ holds because $Q_y e^{-t\Delta_y}$ has a smooth kernel, and $\Theta 2$ is obtained by writing

$$\theta_{b,y}(t) = \mathrm{tr}(e^{-t\Delta_y}) - \mathrm{tr}(P_y e^{-t\Delta_y}).$$

The term $\mathrm{tr}(e^{-t\Delta_y})$ satisfies $\Theta 2$ by §V. 4.2. On the other hand

$$\mathrm{tr}(P_y e^{-t\Delta_y}) = \sum_{k\geq 0} \frac{(-1)^k}{k!} \mathrm{tr}(P_y \Delta_y^k) t^k,$$

so the asymptotic development in that case is in fact a convergent power series. In particular, there is no term involving a negative power of $t$.

By formula (11) from §V.1, we get

$$\zeta'_{>b,y}(0) = \int_1^\infty \theta_{b,y}(t) \frac{dt}{t} + \int_0^1 \rho_{0,y}(t) \frac{dt}{t} + \sum_{i<0} \frac{a_i(y)}{i} + \gamma a_0(y).$$

According to §V.4.2, the coefficients $a_i(y)$ are smooth with respect to $y$.

**3.6.2**    Let us show that $\int_0^1 \rho_{0,y}(t) \frac{dt}{t}$ depends smoothly on $y$. For this it is sufficient to see that, for all $\ell \geq 0$,

(12)
$$\left(\frac{\partial}{\partial y}\right)^\ell \rho_{0,y}(t) = O(t)$$

where $O(t)$ depends on $\ell$ but not on $y$ (and $\left(\frac{\partial}{\partial y}\right)^\ell$ denotes the composite of $\ell$ arbitrary partial derivatives on $Y$). Since $Q_y = 1 - P_y$, the left-hand side of (12) can be written as the difference of two functions of $t$. The first one comes from

$$\left(\frac{\partial}{\partial y}\right)^{\ell} \operatorname{tr}(e^{-t\Delta_v}),$$

and a bound for it follows from

$$(13) \qquad \left\| \left(\frac{\partial}{\partial y}\right)^{\ell} k_t(x, x', y) - \left(\frac{\partial}{\partial y}\right)^{\ell} k_t^N(x, x', y) \right\|_{\sup} \leq C \, t^{N-n},$$

uniformly in $y$, where $k_t^N$ is the approximate kernel and $k_t$ the kernel of $e^{-t\Delta_v}$. This estimate is obtained by a direct extension of the methods of §V.3.5, Step V; see [BGV], proof of Theorem 2.54, p. 93, and [Gr].

The second term in (12) comes from

$$\left(\frac{\partial}{\partial y}\right)^{\ell} \operatorname{tr}(P_y e^{-t\Delta_v}).$$

Let $N = \dim K_y^{q,b} = \operatorname{tr}(P_y)$ and $s_k(y) = \operatorname{tr}(P_y \Delta_y^k)$. The function $s_k(y)$ is smooth in $y$ because $P_y \Delta_y^k$ is an operator with smooth kernel depending smoothly on $y$. We define $\sigma_1(y), \sigma_2(y), \ldots, \sigma_N(y)$ to be the elementary symmetric functions of the eigenvalues of $P_y \Delta_y$. They are given by universal formulas in terms of $s_1(y), \ldots, s_N(y)$, so they are smooth in $y$. Let

$$C = \sup_{\substack{1 \leq i \leq N \\ 0 \leq j \leq \ell}} \sup_y \left| \left(\frac{\partial}{\partial y}\right)^j \sigma_i(y) \right| + 1$$

and choose $A$ such that the following property $\mathcal{P}_k$ is true for $k \leq N$:

$$\mathcal{P}_k : \sup_y \left| \left(\frac{\partial}{\partial y}\right)^j s_k(y) \right| \leq A \, C^k \, N^k \, k^j, \text{ for any } j = 0, 1, \ldots, \ell.$$

Let us prove that $\mathcal{P}_{k-1}$ implies $\mathcal{P}_k$ for $k > N$. We have

$$s_k(y) = \sigma_1(y)s_{k-1}(y) - \sigma_2(y)s_{k-2}(y) + \cdots + (-1)^{N+1}\sigma_N(y)s_{k-N}(y).$$

Applying the Leibniz rule, we get

$$\left| \left(\frac{\partial}{\partial y}\right)^j s_k(y) \right| \leq N \sum_{m=0}^{j} \binom{j}{m} C \, A \, C^{k-1} \, N^{k-1} \, (k-1)^{j-m}$$

$$\leq A \, C^k \, N^k \, k^j.$$

So $\mathcal{P}_k$ is true for all $k$ by induction on $k$. This implies

$$\left(\frac{\partial}{\partial y}\right)^{\ell} \operatorname{tr}(P_y e^{-t\Delta_v}) = \sum_{k \geq 0} \frac{(-1)^k}{k!} t^k \left(\frac{\partial}{\partial y}\right)^{\ell} s_k(y),$$

taking into account that the convergence is uniform with respect to $y$. So, if we remember from Lemma 4 that $s_0(y) = \operatorname{tr}(P_y)$ is locally constant in $y$, we get that

$$\left|\left(\frac{\partial}{\partial y}\right)^\ell \mathrm{tr}(P_y e^{-t\Delta_y})\right| \leq \sum_{k\geq 1} \frac{t^k}{k!} A\, C^k\, N^k\, k^\ell$$

for $\ell \geq 1$, and this is $O(t)$ (uniformly in $y$).

**3.6.3**    Let us prove that $\int_1^\infty \theta_{y,b}(t)\frac{dt}{t}$ depends smoothly on $y$.

To obtain this it will be sufficient to prove that, for every $t \geq 2$ and every $\ell \geq 0$,

$$(14) \qquad\qquad \left|\left(\frac{\partial}{\partial y}\right)^\ell \theta_{y,b}(t)\right| \leq C\, e^{-bt},$$

where $C$ does not depend on $y$. Let $\eta$ be a smooth function on $\mathbf{R}$ such that $0 \leq \eta(s) \leq 1$ , $\eta(s) = 0$ if $s^2 \leq b$ and $\eta(s) = 1$ if $s^2 \geq b+\epsilon$. Define

$$\psi_t(s) = \eta(s) e^{-(t-1)s^2}.$$

This is a function of Schwartz class for $t > 1$. As $\mathrm{Sp}(\Delta_y)\cap[b-\epsilon, b+\epsilon] = \emptyset$, we compute

$$Q_y e^{-t\Delta_y} = \int_{\mathbf{R}} \hat{\psi}_t(\sigma) e^{i\sigma D_y - \Delta_y}\, d\sigma,$$

where the convergence and smoothness is given by the proof of Lemma 2, which applies to any $\psi(s)$ in the Schwartz class. This gives

$$\theta_{y,b}(t) = \int_{\mathbf{R}} \hat{\psi}_t(\sigma)\mathrm{tr}(e^{i\sigma D_y - \Delta_y})\, d\sigma.$$

As in 2.2.2 we get

$$(15) \qquad\qquad \left|\left(\frac{\partial}{\partial y}\right)^\ell \theta_{y,b}(t)\right| \leq C \int_{\mathbf{R}} |\hat{\psi}_t(\sigma)|(1+|\sigma|^\ell)\, d\sigma,$$

where $C$ does not depend on $y$, and does not depend on $t$; notice that all constants arising in the proof of Lemma 2 are independent of the function $\psi(s)$. Now

$$(i\sigma)^\ell \hat{\psi}_t(\sigma) = \frac{1}{2\pi}\int_{\mathbf{R}} e^{-i\sigma s}\left(\frac{d}{ds}\right)^\ell (\psi_t(s))\, ds,$$

with

$$\left|\left(\frac{d}{ds}\right)^\ell (\psi_t(s))\right| \leq C\, t^\ell\, |s|^\ell e^{-(t-1)s^2} \text{ if } s^2 \geq b$$

and

$$\left(\frac{d}{ds}\right)^\ell (\psi_t(s)) = 0 \text{ if } s^2 \leq b.$$

So

$$|\sigma|^\ell |\hat{\psi}_t(\sigma)| \leq \frac{1}{\pi}C\, t^\ell \int_{\sqrt{b}}^\infty s^\ell\, e^{-(t-1)s^2}\, ds \leq C\, t^\ell \int_b^\infty u^{(\ell/2)+1}\, e^{-(t-1)u}\, du.$$

This gives, with another constant $C$,

$$(16) \qquad\qquad |\sigma|^{\ell}|\hat{\psi}_t(\sigma)| \le C \ t^{\ell} \ e^{-(t-1)b}.$$

Combining (15) and (16) (with $\ell + 2$ instead of $\ell$), we get (14). This concludes the proof of the smoothness of $\zeta'_{>b,y}(0)$ with respect to $y$. As $\det_{>b} \Delta$ is equal by definition to $\exp(-\zeta'_{>b,y}(0))$ the proof of Theorem 3 is now complete.                                                                 $\square$

**3.6.4**   As we wrote in the Introduction, the problem of finding a $p$-adic analog for hermitian metrics is still open. In particular Quillen's metric might have a $p$-adic counterpart. Dwork's study of the characteristic power series of the Frobenius endomorphism, viewed as a $p$-adic compact operator [Dw], suggests that this problem might be of arithmetic interest.

# VII

## The Curvature of the Determinant
## Line Bundle

We shall now prove Theorem VI.4 about the curvature of the Quillen
metric on the determinant of the cohomology. This result was obtained
in [BGS1] after special cases had been proved in the relative dimension
one case by Quillen [Q2] and Bismut–Freed [BF]. In fact, for families
of curves, this kind of formula was already familiar to physicists, in
particular in the Polyakov approach to string theory; see for instance
[BK].

A general principle underlying this proof, and also the arithmetic
Riemann–Roch–Grothendieck theorem (Theorem VIII.2), is the following. There are three kinds of secondary objects which enter Arakelov
geometry: Green currents, Bott–Chern classes, and analytic torsion.
But further developments show that these are very similar, and lead to
common generalizations. For instance, when proving Theorem VI.4, we
shall see that the Ray–Singer analytic torsion is really the Bott–Chern
character class of the relative Dolbeault complex. In [BGS2] Green currents and Bott–Chern classes get related, and in [BL] the three objects
are simultaneously generalized.

A basic tool for understanding analytic torsion is the concept of superconnection. Following Quillen [Q3] we explain in §1 how it can be
used to give new representatives of the Chern character. Similarly, in
§2, we give another definition of the Bott–Chern character class of an
acyclic complex of hermitian vector bundles. In §3, under the hypotheses of Theorem VI.4, we use superconnections to define a Bott–Chern
character class of the relative Dolbeault complex, viewed as a complex of

infinite-dimensional bundles on the base. Using the local index theorem for families [B1] (see also [BV] and [BGV]) and a result of Berline–Vergne [BV], we then prove Theorem VI.4.

## 1. Superconnections

**1.1**    Let $M$ be a smooth manifold and $E = E^+ \oplus E^-$ a $\mathbb{Z}_2$-graded vector bundle on $M$; $E^+$ is called the even part of $E$, and $E^-$ the odd part. Equivalently, $E$ is equipped with an involution $\epsilon \in \operatorname{End}(E)$, $\epsilon^2 = \operatorname{id}$, and

$$E^+ = \ker(\epsilon - \operatorname{id}),$$
$$E^- = \ker(\epsilon + \operatorname{id}).$$

There is then a canonical involution on $\operatorname{End}(E)$, sending $\alpha$ to $\epsilon\alpha\epsilon$. So $\operatorname{End}(E)$ is also $\mathbb{Z}_2$-graded, namely

$$\operatorname{End}(E)^+ = \operatorname{Hom}(E^+, E^+) \oplus \operatorname{Hom}(E^-, E^-),$$
$$\operatorname{End}(E)^- = \operatorname{Hom}(E^+, E^-) \oplus \operatorname{Hom}(E^-, E^+).$$

The *supercommutator* of $\alpha, \beta \in \operatorname{End}(E)$ is

$$[\alpha, \beta] = \alpha\beta - (-1)^{|\alpha||\beta|}\beta\alpha,$$

where $|\alpha| = 0$ if $\alpha \in \operatorname{End}(E)^+$, and $|\alpha| = 1$ if $\alpha \in \operatorname{End}(E)^-$; the formula is extended by linearity to non-homogeneous $\alpha$ and $\beta$ . It satisfies the (super) Leibniz rule:

$$[\alpha, \beta\gamma] = [\alpha, \beta]\gamma + (-1)^{|\alpha||\beta|}\beta[\alpha, \gamma];$$

in general, each time one permutes an element of degree $n$ with an element of degree $m$, the classical formula has to be modified by a sign $(-1)^{nm}$. From now on [ , ] will always stand for a supercommutator. Finally, a linear functional called the *supertrace* is defined on $\operatorname{End}(E)$ by

$$\operatorname{tr}_s(\alpha) := \operatorname{tr}(\epsilon\alpha).$$

It vanishes on odd elements and on supercommutators.

**1.2**    Consider the left $A^{\cdot}(M)$-module

$$V = A^{\cdot}(M) \otimes_{C^\infty(M)} \Gamma(M, E).$$

We can view $V$ as a right $A^{\cdot}(M)$-module by

$$\alpha\,\omega := (-1)^{|\alpha||\omega|}\omega\,\alpha;$$

notice that $A^{\cdot}(M)$ is a commutative superalgebra : $\omega\omega' = (-1)^{|\omega||\omega'|}\omega'\omega$. The module $V$ has a natural $\mathbb{Z}/2$ grading which combines the $\mathbb{Z}$-grading of $A^{\cdot}(M)$ and the $\mathbb{Z}/2$-grading of $\Gamma(M, E)$.

Let $\nabla$ be a connection on $E$ with $\nabla(\epsilon) = 0$; equivalently $\nabla = \nabla^+ + \nabla^-$ with $\nabla^+$ a connection on $E^+$ and $\nabla^-$ a connection on $E^-$. Let $u \in \text{End}(E)^-$. The operator $A = \nabla^+ + \nabla^- + u$ is called a *superconnection*. It maps $\Gamma(M, E)$ to $\Gamma(M, E) \oplus A^1(M) \otimes_{C^\infty(M)} \Gamma(M, E)$. We extend this action of $A$ to $V$ by the formula

$$A(\omega s) := d\omega.s + (-1)^{|\omega|}(\omega \wedge \nabla s + \omega u(s)).$$

As is easily checked, this implies

(1)    $$A(\omega \wedge v) = d\omega \wedge v + (-1)^{|\omega|}\omega \wedge Av$$

for any $\omega \in A^{\cdot}(M)$, $v \in V$.

**Definition 1**    (see [Q3]) *A superconnection on $E$ is an odd degree element of $\text{End}_{\mathbb{C}}(V)$ satisfying (1).*

We also have $A(v \wedge \omega) = (Av) \wedge \omega + (-1)^{|v|}v \wedge d\omega$. Furthermore (1) implies, as in §IV.2.1,

$$A^2(\omega \wedge v) = \omega \wedge A^2(v).$$

As $A^2$ is even, this may as well be written $[A^2, \omega] = 0$. So $A^2$ lies in the algebra

$$\mathcal{A} = \{\alpha \in \text{End}_{\mathbb{C}}(V)/ \text{ for all } \omega \in A^{\cdot}(M), [\alpha, \omega] = 0\}.$$

This algebra may also be described as follows (see [Q3]):

(2)    $$\mathcal{A} = A^{\cdot}(M) \hat{\otimes}_{C^\infty(M)} \Gamma(M, \text{End}(E)),$$

where the tensor product is taken in the sense of superalgebras:

$$\omega \otimes \alpha = (-1)^{|\alpha||\omega|}\alpha \otimes \omega.$$

**1.3**

**Proposition 1**    *For any superconnection $A$ on $E$, the form $\text{tr}_s \exp(-A^2)$ is closed on $M$.*

*Proof.* First, this form is well-defined because $A^2 \in \mathcal{A}$, and by (2) there is a natural extension of $\text{tr}_s$ to $\mathcal{A}$ with values in $A^{\cdot}(M)$.

Now, given any two superconnections $A, A'$ then

$$(A - A')(\omega v) = (-1)^{|\omega|}\omega \wedge (A - A')(v)$$

so $A - A'$ is an odd element of $\mathcal{A}$. Since the statement we want to prove is local, after trivializing $E$, we may write $A = d + \theta$ with $\theta$ an odd element of $\mathcal{A}$. Then

$$d \, \text{tr}_s \exp(-A^2) = \text{tr}_s d(\exp(-A^2))$$
$$= \text{tr}_s[d, \exp(-A^2)]$$
$$= \text{tr}_s[A, \exp(-A^2)] - \text{tr}_s[\theta, \exp(-A^2)].$$

The first term is zero because $A$ (super)commutes with $A^2$. The second term is zero because of the next lemma.

**Lemma 1**   *For any $a, b$ in $\mathcal{A}$,   $\mathrm{tr}_s[a, b] = 0$.*

*Proof.*   Writing $a = \alpha\omega, b = \beta\omega'$ , $\alpha, \beta \in \Gamma(M, \mathrm{End}(E))$, and using the super Leibniz rule, we have

$$[a, b] = [a, \beta\omega'] = [a, \beta]\omega'$$

because $[a, \omega'] = 0$, therefore

$$[a, b] = [\alpha\omega, \beta]\omega' = (-1)^{|\omega||\beta|}[\alpha, \beta]\omega\omega'$$

since $[\omega, \beta] = 0$. So

$$\mathrm{tr}_s[a, b] = (-1)^{|\omega||\beta|}(\mathrm{tr}_s[\alpha, \beta])\omega\omega' = 0.$$

$\square$

It is shown in [Q3] that the cohomology class of $\mathrm{tr}_s \exp(-A^2)$ is, up to powers of $2\pi i$, the Chern character of $E^+$ minus the Chern character of $E^-$.

## 2. Bott–Chern classes via superconnections

Superconnections provides new definitions of the Bott–Chern secondary characteristic classes which we introduced in §IV.3. More generally, let

$$E_{\cdot} : 0 \to E_0 \xrightarrow{v} E_1 \xrightarrow{\quad} \cdots \xrightarrow{v} E_n \longrightarrow 0$$

be an *acyclic* complex of holomorphic vector bundles on a complex manifold $X$. Assume that each $E_i$ is equipped with an hermitian metric $h_i$. Let $E^+ = \bigoplus_p E_{2p}$, $E^- = \bigoplus_p E_{2p+1}$, and $E = E^+ \oplus E^-$, and endow these bundles with the orthogonal direct sum of the $h_i$'s.

These metrics give hermitian holomorphic connections $\nabla^+$ (resp. $\nabla^-$) on $E^+$ ( resp. $E^-$); see Lemma IV.1. Also $v$ acts as an odd endomorphism of $E$. Write $v^*$ for the adjoint of $v$. Therefore

$$A = \nabla^+ + \nabla^- + v + v^*$$

is a superconnection on $E_{\cdot}$.

As $E_{\cdot}$ is an acyclic complex and since the Chern character is additive in cohomology, we know that $\sum_{i \geq 0}(-1)^i\mathrm{ch}(E, h_i)$ is exact on $X$. In the following we shall achieve a *double transgresssion* and express this form as the image by $dd^c$ of an explicit element.

We first pullback $(E_{\cdot}, h)$ to a complex with metrics on $X \times \mathbb{C}$. Then, writing $\delta = dz\frac{\partial}{\partial z} + d\bar{z}\frac{\partial}{\partial \bar{z}}$ and $\nabla = \nabla^+ + \nabla^-$, we consider

$$A_z = \nabla + \delta + zv + \bar{z}v^*,$$

which is a superconnection on $E$ (over $X \times \mathbb{C}$). Let

$$\eta = \int_{\mathbb{C}} \text{tr}_s \exp(-A_z^2) \log|z|^2,$$

i.e. the integral of the $dz\,d\bar{z}$ part of $\text{tr}_s \exp(-A_z^2)$ against $\log|z|^2$.

Let $*$ be the operator acting by multiplication by $(\frac{1}{2\pi i})^p$ on $A^{p,p}(M)$.

**Proposition 2**    $(dd^c \eta)^* = \sum_{i \geq 0} (-1)^i \text{ch}(E_i, h_i)$.

*Proof.* First we investigate the convergence of $\eta$. Notice that

$$A_z^2 = |z|^2 (vv^* + v^* v) + R_z$$

where $R_z$ has form-degree $\geq 1$ (on $X \times \mathbb{C}$), and so $R_z$ is nilpotent. Thanks to Duhamel's formula (Chapter V, (15)) we have

$$\exp(-A_z^2) - \exp(|z|^2 \Delta) =$$

$$\sum_{k \geq 1} (-1)^k \int_{\Delta^k} e^{-(1-t_k)|z|^2 \Delta} R_z \cdots R_z e^{-t_1 |z|^2 \Delta} dt_1 \cdots dt_k$$

where $\Delta = vv^* + v^*v$ and the sum is finite ($k \leq n = \dim_{\mathbb{C}}(X)$). As $E$. is acyclic, $\Delta$ has a smallest eigenvalue $\lambda > 0$. Writing $\|\cdot\|$ for the operator norm, we get

$$\|e^{-(t_j - t_{j-1})|z|^2 \Delta}\| \leq 1 \qquad\qquad \text{for all } j,$$

$$\|e^{-(t_j - t_{j-1})|z|^2 \Delta}\| \leq e^{-\lambda \frac{1}{k}|z|^2} \qquad\qquad \text{for at least one } j.$$

As $R_z = \nabla^2 + v\,dz + v^*\overline{dz} + z\nabla(v) + \bar{z}\nabla(v^*)$ is linear in $z, \bar{z}$, we get:

$$\|\exp(-A_z^2)\| \leq C(1 + |z|^n) e^{-\frac{\lambda}{n}|z|^2},$$

where $C$ is uniform on $X$.

Similar estimates hold for any number of derivatives of $\exp(-A_z^2)$ with respect to $z, \bar{z}$, uniformly with respect to $X$. So we may extend $a(z) = \text{tr}_s \exp(-A_z^2)$ in a smooth way to $X \times \mathbb{P}^1$ by declaring it to be zero on the fiber $X \times \{\infty\}$. As it is closed on the dense open set $X \times \mathbb{C}$ (Proposition 1), it is also closed on $X \times \mathbb{P}^1$.

We will now make use of the fact, which we do not prove here (but see Proposition 3 below for a proof in a slightly more complicated case), that $a(z) \in \bigoplus_{p \geq 0} A^{p,p}(X \times \mathbb{P}^1)$. This implies that $a(z)$ is not only $d_{X \times \mathbb{P}^1}$-closed but also $d^c_{X \times \mathbb{P}^1}$-closed. Therefore

$$d_X d^c_X \int_{\mathbb{P}^1} a(z) \log|z|^2 = \int_{\mathbb{P}^1} d_{X \times \mathbb{P}^1} d^c_{X \times \mathbb{P}^1} (a(z) \log|z|^2)$$

$$= \int_{\mathbb{P}^1} a(z) \delta\delta^c \log|z|^2$$

$$= i_0^* a(z) - i_\infty^* a(z) = i_0^* a(z)$$

(we used $\delta\delta^c \log|z|^2 = \delta_0 - \delta_\infty$).

Finally,   $A_0^2 = \nabla^2 + v\,dz + v^*\overline{dz}$, so $i_0^*a(z) = \mathrm{tr}_s \exp(-\nabla^2)$. Therefore
$$(dd^c\eta)^* = \mathrm{tr}_s \exp(-\nabla^2)^* = \sum_{i \geq 0}(-1)^i \mathrm{ch}(E_i, h_i).$$

$\square$

## 3. The Bott–Chern character class of the relative Dolbeault complex

**3.1**    Remember that our intention is to prove Theorem VI.4. Using Theorem VI.5, we see that is sufficient to prove Theorem VI.4 for *one* choice of the metric $h_f$. Using the "locally Kähler" hypothesis we are then reduced to the situation where $X$ itself has a Kähler metric, and where $h_f$ is the restriction of this Kähler metric to $Tf$. (We shall not prove Theorem VI.5 in these notes, but Theorem VI.4 is already of interest under the assumption we just made.)

On $Y$, we have a complex of infinite-dimensional vector bundles
$$0 \to D^0 \xrightarrow{\overline{\partial}} D^1 \xrightarrow{\overline{\partial}} \cdots \xrightarrow{\overline{\partial}} D^n \longrightarrow 0\,,$$
where $D^q(U) = C^\infty(f^{-1}(U),\ \Lambda^q(Tf^{*0,1}) \otimes E)$ and $\overline{\partial} = \overline{\partial}_f$.
   In order to define a connection on $D = \bigoplus_{q \geq 0} D^q$, we use the orthogonal splitting
$$TX_x = Tf_x \oplus (TX_x)^H$$
given by the metric on $X$; vectors in $(TX_x)^H$ are called horizontal. The map $f$ induces an isomorphism $d_x f : (TX_x)^H \xrightarrow{\sim} T_{f(x)}Y$ for every point $x \in X$. Therefore any vector field $v$ on $Y$ has an horizontal lift $v^H$ on $X$.
   Let $\nabla$ be the hermitian holomorphic connection on $(E, h)$ over $X$. Define, for $v \in TY$ and $\sigma \in D(Y)$,
$$(3) \qquad\qquad \tilde{\nabla}_v(\sigma) = \nabla_{v^H}(\sigma).$$
One checks that, for any $\varphi \in C^\infty(Y)$,
$$\tilde{\nabla}_{\varphi v}(\sigma) = \varphi\,\tilde{\nabla}_v(\sigma)$$
and
$$\tilde{\nabla}_v(\varphi\sigma) = v(\varphi)\sigma + \varphi\,\tilde{\nabla}_v(\sigma).$$
   So we get a connection on $D$
$$\tilde{\nabla} : D \longrightarrow A^1(Y, D).$$
As usual, $\tilde{\nabla}$ is extended by the Leibniz rule to act upon $\bigoplus_j A^j(Y, D)$.

**Lemma 2**    $\tilde{\nabla}$ *is unitary for the $L^2$-metric on $D$.*

*Proof.* Let $\mu$ be the volume form on $X_y$. For $v \in T_y Y$ and $s, t$ sections of $D$ in a neighborhood of $y$, we compute the Lie derivative:

$$d_v \langle s, t \rangle = d_v \left( \int_{X_y} \langle s(x), t(x) \rangle \mu \right)$$

$$= \int_{X_y} d_{v^H} \langle s(x), t(x) \rangle \mu + \int_{X_y} \langle s(x), t(x) \rangle d_{v^H} \mu.$$

Now

$$d_{v^H} \langle s(x), t(x) \rangle = \langle \nabla_{v^H} s(x), t(x) \rangle + \langle s(x), \nabla_{v^H} t(x) \rangle$$

so that

$$\int_{X_y} d_{v^H} \langle s(x), t(x) \rangle \mu = \langle \tilde{\nabla}_v s, t \rangle + \langle s, \tilde{\nabla}_v t \rangle.$$

So it will be enough to prove $d_{v^H}(\mu) = 0$.

Trivializing $Tf$ near $x \in X_y$, we have, up to some constant factor,

$$\mu = [\det(g)]^{-1/2} \prod_i dz_i \overline{dz_i}$$

where $g$ is the matrix giving the Riemannian metric on $Tf$. So we want to prove $d_{v^H}(\det g) = 0$, and for this it is sufficient to prove $\text{tr} d_{v^H} g = 0$ because

$$\text{tr} d_{v^H} g = d_{v^H}(\log \, \det g) = \frac{1}{\det g} d_{v^H}(\det g).$$

Let $(e_i)$ be an orthonormal frame in $(Tf)^{1,0}$, and $(\bar{e}_i)$ the conjugate frame of $(Tf)^{0,1}$. As $\langle e_i, e_i \rangle = 1$ identically, we have

$$(4) \qquad 0 = d_{v^H} \langle e_i, e_i \rangle = d_{v^H}(g)(e_i, \bar{e}_i) + \langle [v^H, e_i], e_i \rangle + \langle e_i, \overline{[v^H, e_i]} \rangle.$$

Let $\omega$ be the (normalized) Kähler form on $X$:

$$\omega(u, v) = \frac{i}{2\pi} \langle u, \bar{v} \rangle.$$

The Kähler condition $d\omega = 0$ gives, for $u, v, t$ any vector fields

$$(5)$$
$$0 = t\omega(u, v) + u\omega(v, t) + v\omega(t, u) + \omega([t, u], v) + \omega([u, v], t) + \omega([v, t], u).$$

We write this for $t = v^H$, $u = e_i$, $v = \bar{e}_i$ and observe that

$$v^H \omega(e_i, \bar{e}_i) = \frac{i}{2\pi} v^H \langle e_i, e_i \rangle = 0.$$

Furthermore

$$e_i \omega(v^H, \bar{e}_i) = \frac{i}{2\pi} e_i \langle v^H, e_i \rangle = 0$$

and

$$\bar{e}_i \omega(v^H, e_i) = 0$$

because the horizontal space is the orthogonal complement of $Tf$. Finally

$$\omega([e_i, \bar{e}_i], v^H) = 0$$

because $[e_i, \bar{e}_i]$ is vertical. So we find from (5)

$$\omega([v^H, e_i], \bar{e}_i) + \omega([\bar{e}_i, v^H], e_i) = 0.$$

As $\omega([\bar{e}_i, v^H], e_i) = \omega(e_i, [v^H, \bar{e}_i])$, this gives

$$\langle [v^H, e_i], e_i \rangle + \langle e_i, \overline{[v^H, \bar{e}_i]} \rangle = 0.$$

Now $\overline{[v^H, \bar{e}_i]} = [v^H, e_i]$. Restricting to real vectors and comparing to (4) above, we get

$$d_{v^H}(g)(e_i, e_i) = 0,$$

so $tr\, d_{v^H}(g) = 0$ and we are done.    □

**3.2**    Mimicking the finite-dimensional situation of §2, we introduce the following superconnection on the pull-back of $D$ to $Y \times \mathbb{C}^*$ :

$$B_z = \tilde{\nabla} + \delta + z\bar{\partial} + \bar{z}\,\bar{\partial}^*.$$

**Lemma 3**    *The operator* $\exp(-B_z^2)$ *is trace-class.*

*Proof.* We may write $B_z^2 = |z|^2 \Delta + \Phi$ where $\Phi$ has degree $\geq 1$ and is therefore nilpotent. So $\exp(-B_z^2)$ may be defined as the operator with smooth kernel given by Duhamel's formula (which is a *finite* sum for degree reasons):

$$\exp(-B_z^2) =$$

$$\exp(-|z|^2 \Delta) + \sum_{k \geq 1} (-1)^k \int_{\Delta^k} e^{-(1-t_k)|z|^2 \Delta}.\Phi. \cdots .\Phi. e^{-t_1|z|^2 \Delta} dt_1 \cdots dt_k.$$

In particular it is trace-class.    □

Let

$$b(z) = \mathrm{tr}_s\, \exp(-B_z^2).$$

**Proposition 3**    *The form* $b(z)$ *lies in* $\bigoplus_{p \geq 0} A^{p,p}(Y \times \mathbb{C}^*)$. *It is closed and invariant under rotation.*

*Proof.* If $\theta \in \mathbb{R}$ and $(y, z) \in Y \times \mathbb{C}^*$ define $r_\theta(y, z) = (y, e^{i\theta}z)$. When saying that $b(z)$ is invariant under rotation we mean that $r_\theta^*(b(z)) = b(z)$ for every $\theta \in \mathbb{R}$.

The fact that $b(z)$ is closed is shown as in the finite-dimensional case (Proposition 1). On the other hand, let $\mathcal{B}$ be the subalgebra of

$$\mathcal{A} = \Gamma(Y \times \mathbb{C}^*, \mathrm{End}(D)) \hat{\otimes} \left( \bigoplus_{p,q} A^{p,q}(Y \times \mathbb{C}^*) \right)$$

generated as a vector space by forms $\alpha$ of degree $(n, p, q) \in \mathbb{Z} \times \mathbb{N} \times \mathbb{N}$ such that $p = q + n$ and $r_\theta^* \alpha = e^{in\theta} \alpha$, where $n$ is the degree coming from $\mathrm{End}(D)$.

**Lemma 4**     $B_z^2$ *lies in* $\mathcal{B}$.

*Proof of the Lemma.* We compute

$$
(6) \qquad
\begin{aligned}
B_z^2 &= (\tilde{\nabla} + \delta + z\overline{\partial} + \overline{z}\overline{\partial}^*)^2 \\
&= \tilde{\nabla}^2 + |z|^2 \Delta + z\tilde{\nabla}(\overline{\partial}) + \overline{z}\,\tilde{\nabla}(\overline{\partial}^*) + dz\,\overline{\partial} + \overline{dz}\,\overline{\partial}^*
\end{aligned}
$$

and prove that each term in this sum lies in $\mathcal{B}$:

- Let $\tilde{\nabla} = \tilde{\nabla}^{1,0} + \tilde{\nabla}^{0,1}$. If $v \in T^{0,1}Y$ then $v^H \in T^{0,1}X$, and since $\nabla^{0,1} = \overline{\partial}_E$, we get $(\tilde{\nabla}^{0,1})^2 = 0$. As $\tilde{\nabla}$ is unitary this implies $(\tilde{\nabla}^{1,0})^2 = 0$. So $\tilde{\nabla}^2$ is of type $(0, 1, 1)$. Being independent of $z$ it is also rotation invariant. So it lies in $\mathcal{B}$.
- $|z|^2 \Delta$ is of type $(0, 0, 0)$ and invariant by rotation.
- $z\,\tilde{\nabla}(\overline{\partial}) = z\,\tilde{\nabla}^{1,0}(\overline{\partial})$ has type $(1, 1, 0)$ and clearly $r_\theta^* z = e^{i\theta} z$, so it lies in $\mathcal{B}$.
- $z\,\tilde{\nabla}(\overline{\partial}^*) = z\,\tilde{\nabla}^{0,1}(\overline{\partial}^*)$ by unitarity and the above, so it lies in $\mathcal{B}$; it is of type $(-1, 0, 1)$.
- $dz\,\overline{\partial}$ is of type $(1, 1, 0)$ and clearly $r_\theta^* dz = e^{i\theta} dz$ so it lies in $\mathcal{B}$. The same argument applies to $\overline{dz}\,\overline{\partial}^*$.

$\square$

Now, to prove Proposition 3, notice that $\mathrm{tr}_s(\alpha) = 0$ if $\alpha$ is of type $(n, p, q)$ with $n \neq 0$. Therefore, if $\alpha \in \mathcal{B}$, then $\mathrm{tr}_s \alpha \in \bigoplus_p A^{p,p}(Y \times \mathbb{C}^*)$ and it is rotation invariant. According to Lemma 4, this is the case for $\mathrm{tr}_s \exp(-B_z^2)$. This proves Proposition 3.     $\square$

**3.3**     Let us introduce, when $0 < t < T$ are two real numbers,

$$
\beta(t, T) = \int_{t \leq |z| \leq T} b(z)\, \log |z|^2.
$$

The form $b(z)$ is $(d_Y + \delta)$-closed and it lies in $\bigoplus_p A^{p,p}(Y \times \mathbb{C}^*)$, so it is also $(d_Y^c + \delta^c)$-closed. Hence

$$
d_Y d_Y^c b(z) = -d_Y \delta^c b(z) = \delta^c d_Y b(z) = -\delta^c \delta b(z) = \delta \delta^c b(z).
$$

We deduce that

$$
\begin{aligned}
d_Y d_Y^c \beta(t, T) &= \int_{t \leq |z| \leq T} (d_Y d_Y^c b(z))\, \log |z|^2 \\
&= \int_{t \leq |z| \leq T} (\delta \delta^c\, b(z))\, \log |z|^2.
\end{aligned}
$$

Let us apply the Stokes formula to this integral:

$$d_Y d_Y^c \beta(t,T) = \int_{t \le |z| \le T} (\delta^c b(z)) \, \delta \log |z|^2$$
$$+ \int_{|z|=T} \delta^c b(z) \log |z|^2 - \int_{|z|=t} (\delta^c b(z)) \log |z|^2.$$

Using $\delta^c \delta \log |z|^2 = 0$ we obtain

$$d_Y d_Y^c \beta(t,T) = \int_{|z|=t} b(z) \delta^c \log |z|^2 - \int_{|z|=T} b(z) \delta^c \log |z|^2$$
$$- \int_{|z|=t} (\delta^c b(z)) \log |z|^2 + \int_{|z|=T} (\delta^c b(z)) \log |z|^2.$$

If we use polar coordinates $z = re^{i\theta}$, we have $4\pi \delta^c = r\frac{\partial}{\partial r} d\theta - \frac{1}{r}\frac{\partial}{\partial \theta} dr$, hence $\delta^c \log |z|^2 = \frac{d\theta}{2\pi}$. Since $b(z)$ is invariant under rotation, we get finally, with $i_z(y) = (y, z)$,

(7)
$$d_Y d_Y^c \beta(t,T) = i_t^* b(z) - i_T^* b(z) - t \, \log(t) \, i_t^* \left(\frac{\partial b}{\partial r}\right) + T \, \log(T) \, i_T^* \left(\frac{\partial b}{\partial r}\right).$$

To proceed further we need to know the behavior of $b(z)$ when $|z| \to 0$ and $+\infty$.

## 4. The family index theorem

**4.1**    To study the behavior of $b(z)$ when $|z| \to 0$ we use the following theorem:

**Theorem 1**    (Bismut, [B1], [BV], [BGV])

$$\lim_{t \to 0} \left[ \mathrm{tr}_s \exp(-\tilde{\nabla} + t \, D_y)^* \right]^{(\le 2)} = \left[ \int_{X/Y} \mathrm{ch}(E, h) Td(f) \right]^{(\le 2)}.$$

Here $D_y = \sqrt{2}(\bar\partial + \bar\partial^*)$ is the Dirac operator along the fiber, and $[\alpha]^{(\le 2)}$ is the sum of the components of degree at most two with respect to $Y$ of the form $\alpha$. This equality may be rewritten

(8)        $$\lim_{t \to 0} [i_t^* b(z)^*]^{(\le 2)} = \left[ \int_{X/Y} \mathrm{ch}(E, h) Td(f) \right]^{(\le 2)}.$$

**Remarks**
    The proof in [B1] is actually given for arbitrary Dirac operators (on the right-hand side of the formula one uses the $\hat{A}$ genus instead of the Todd genus). Furthermore it is valid in all degrees if one adds to $\tilde{\nabla} + t \, D_y$ a

*counterterm* $\frac{1}{4t}c(T)$, where $c(T)$ is Clifford multiplication by some tensor coming from the fact that the horizontal tangent bundle is not integrable. Since this counterterm has degree bigger than two with respect to $Y$, it does not affect the formula (8). Note also that there is a *local* equality of forms on $X$ from which Theorem 1 follows by integration along the fibers of $f$.

When $Y$ is a point, we get the theorem of Riemann-Roch-Hirzebruch [Hz]. Indeed, for all $t > 0$,

$$\text{tr}_s \exp(-t\Delta) = \chi(E)$$

is the Euler characteristic of $E$.

**4.2**    The behavior of $b(z)$ when $|z| \to \infty$ is given by the following theorem of Berline–Vergne. Let $f : M \to B$ be a smooth proper map of $C^\infty$-manifolds, $E = E^+ \oplus E^-$ a super-vector bundle on $M$ with hermitian metric $h$, $h_f$ a Riemannian metric on $Tf$, and $u_y \in \text{End}(V_y)$ a smooth family of odd, elliptic, formally self-adjoint, first-order differential operators on $V_y = \Gamma(f^{-1}(y), E)$. Choose a connection $\nabla$ on $E$ and define $\tilde{\nabla}$ on $\bigoplus_j A^j(B, V)$ as we did above in §3.1. Finally, let $\delta_t$ be the scaling operator defined by $\delta_t = t^{-n/2}$ on $A^n(B, \text{End}(V))$   $(t > 0)$.

**Theorem 2**    (Berline–Vergne,[BV], [BGV])*Assume that* $\dim(\ker(u_y))$ *is constant in $y$. Then, as $t \to \infty$,*

$$\text{tr}_s \delta_t \exp(-tA^2) = \text{tr}_s \exp(-(P\tilde{\nabla})^2) + O(\frac{1}{\sqrt{t}}),$$

*where $A$ is the superconnection $\tilde{\nabla} + u_y$ and $P$ is the orthogonal projection onto* $\ker u_y$.

The limit in Theorem 2 is for any $C^k$-norm on $B$ and the rest $O(\frac{1}{\sqrt{t}})$ is uniform with respect to $y$ (restricted to a compact set).

**4.3**    We now return to our original situation. We want to apply the Theorem 2 to $f : X \times \mathbb{C}^* \to Y \times \mathbb{C}^*$, with $A = \tilde{\nabla} + \delta + u_{y,z}$ and $u_{y,z} = z\,\bar{\partial} + \bar{z}\,\bar{\partial}^*$.

First, $u_{y,z}^2 = |z|^2\Delta$ and $\ker(u_{y,z})$ is independent of $z$ (and so will be the projection $P$ on this kernel). Notice that it is sufficient to prove the equality of Theorem VI.4 on a dense open subset $U$ of $Y$ since both sides are smooth forms on $Y$. The following lemma shows that this enables us to assume that the kernel of $u_{y,z}$ has a constant dimension.

**Lemma 5**    *There is a dense open subset $U \subset Y$ such that, for every integer $q \geq 0$, the sheaf $R^q f_* E$ is locally free on $U$.*

*Proof.* This is true for any coherent $\mathcal{O}_Y$-module $M$ instead of $R^q f_* E$. In the case where $M$ admits a resolution of length one

$$0 \longrightarrow P_1 \overset{v}{\longrightarrow} P_0 \longrightarrow M \longrightarrow 0$$

by vector bundles, $\operatorname{rank}_y(M) = \operatorname{corank}_y(v)$ is constant outside of a subvariety defined by the vanishing of certain minors of $v$, and on this open set $M$ is locally free. In general, starting from a resolution

$$0 \longrightarrow P_n \overset{v_n}{\longrightarrow} P_{n-1} \longrightarrow \cdots \longrightarrow P_0 \overset{v_0}{\longrightarrow} M \longrightarrow 0$$

with vector bundles, we apply the preceeding argument to $\operatorname{coker} v_n$ and, by induction on $n$, we deduce the result for $M$. $\qquad\qquad\square$

From now on, until the end of this chapter, we shall assume that $R^q f_* E$ is locally free on $Y$. This is made possible by the previous lemma and the discussion before it.

**Lemma 6**     *Decompose* $b(z) = b(z)^{[0]} + b(z)^{[1]} + b(z)^{[2]}$ *according to the degree on* $\mathbb{C}^*$. *Then, uniformly on* $Y$,

$$b(z)^{[0]} = \operatorname{tr}_s \exp(-(P\widetilde{\nabla})^2) + O(\frac{1}{|z|}),$$

$$b(z)^{[1]} = O(\frac{1}{|z|^2}),$$

$$b(z)^{[2]} = O(\frac{1}{|z|^3}),$$

*and*

$$\frac{\partial}{\partial r} b(r)^{[0]} = O(\frac{1}{r^2}).$$

*Proof.* Recall from (6) that

(9) $\qquad B_z^2 = \widetilde{\nabla}^2 + |z|^2 \Delta + z \, \widetilde{\nabla}(\overline{\partial}) + \overline{z} \, \widetilde{\nabla}(\overline{\partial}^*) + dz \, \overline{\partial} + \overline{dz} \, \overline{\partial}^*.$

From this we get

(10)
$$\delta_t(t \, B_z^2) = \widetilde{\nabla}^2 + t|z|^2 \Delta + \sqrt{t} \, z \, \widetilde{\nabla}(\overline{\partial}) + \sqrt{t} \, \overline{z} \, \widetilde{\nabla}(\overline{\partial}^*) + \sqrt{t} \, dz \, \overline{\partial} + \sqrt{t} \, \overline{dz} \, \overline{\partial}^*.$$

Let $\gamma_t$ be the scaling operator on $\mathcal{A}$ sending $dz$ to $\sqrt{t} \, dz$, and $d\overline{z}$ to $\sqrt{t} \, d\overline{z}$. We deduce from the two equalities (9) and (10) above that

(11) $\qquad\qquad\qquad \delta_t(t \, B_z^2) = \gamma_t \, B_{\sqrt{t} \, z}^2;$

in other words $\delta_t(t B_z^2)$ is the pull-back of $B_z^2$ under the map $z \, \to \, \sqrt{t} \, z$.

Writing the equality (11) at the point $z = 1$ and $t = r^2$ we get:

$$\delta_{r^2} \exp(-r^2 B_z^2)_{|z=1} = \gamma_{r^2} \exp(-B_z^2)_{|z=r}.$$

Then Theorem 2, when applied to $Y \times \mathbb{C}^*$ and $A = B_z$, gives

(12) $\qquad \operatorname{tr}_s \delta_{r^2} \exp(-r^2 B_z^2) = \operatorname{tr}_s \exp(-(P\widetilde{\nabla})^2) + O(1/r),$

where $O(1/r)$ is uniform on $U \times S^1$ for any relatively compact open subset $U \subset Y$. Notice that $P\delta = \delta P$ and $\tilde{\nabla}\delta = -\delta\tilde{\nabla}$ so that $(P(\tilde{\nabla} + \delta))^2 = (P\tilde{\nabla})^2$. Restricting $z$ to 1, (11) and (12) become

(13)    $\operatorname{tr}_s \gamma_{r^2} \exp(-B_z^2)|_{z=r} = \operatorname{tr}_s \exp(-(P\tilde{\nabla})^2) + O(1/r)$.

Now $\operatorname{tr}_s \exp(-(P\tilde{\nabla})^2)$ is of degree 0 with respect to $\mathbf{C}^*$. So using the fact that $\gamma_{r^2}$ is $r$ on 1-forms and $r^2$ on $dz\,\overline{dz}$, we obtain the first three equalities of the lemma, since $b(z)$ is invariant under rotation. Here the equality

$$b(z)^{[1]} = O\left(\frac{1}{|z|^2}\right)$$

means that the form $b(z)^{[1]}$ can be written $\alpha dz + \beta d\overline{z}$ with $\alpha = O(\frac{1}{|z|^2})$ and $\beta = O(\frac{1}{|z|^2})$, and similarly for $b(z)^{[2]} = O(\frac{1}{|z|^3})$.

We know by Proposition 3 that $b(z)$ is $d_{Y \times \mathbf{C}^*}$-closed. Moreover $\frac{\partial}{\partial\theta} b(z) = 0$ since $b(z)$ is invariant under rotation. Therefore

(14)    $$\frac{\partial}{\partial r}(b(z))dr = -d_Y\, b(z).$$

By the Remark after Theorem 2, the above proof also gives $d_Y\, b(z)^{[1]} = O(\frac{1}{|z|^2})$. So

$$\frac{\partial}{\partial r}(b(r))^{[0]} = O\left(\frac{1}{r^2}\right).$$

$\square$

**4.4**    Denote by $(R^q f_* E, L^2)$ the bundle $R^q f_* E$ on $Y$, endowed with its $L^2$ metric.

**Lemma 7**    $\operatorname{tr}_s \exp(-(P\tilde{\nabla})^2)^* = \sum_{q \geq 0}(-1)^q \operatorname{ch}(R^q f_* E, L^2)$.

*Proof.* It is sufficient to prove that $P\tilde{\nabla}$ is the hermitian holomorphic connection on $\bigoplus_{q \geq 0} R^q f_* E$ for the $L^2$-metric. Since $\tilde{\nabla}_v^{0,1} = \overline{\partial}_{v^H}$ commutes with $\overline{\partial}_{X_y}$, it preserves $\ker \Delta^q = (R^q f_* E)_y$, and $(P\tilde{\nabla})^{0,1} = \overline{\partial}_{R^q f_* E}$. On the other hand, given $s, t$ in $\ker \Delta^q$ and $v$ in $TY$, we have

$$v\langle s, t \rangle = \langle \tilde{\nabla}_v s, t \rangle_{L^2} + \langle s, \tilde{\nabla}_v t \rangle_{L^2} = \langle P\tilde{\nabla}_v s, t \rangle_{L^2} + \langle s, P\tilde{\nabla}_v t \rangle_{L^2}.$$

This proves that $P\tilde{\nabla}$ is unitary.    $\square$

# 5.  Conclusion

**5.1**    We may now take the limit as $T \to \infty$ and $t \to 0$ in the formula

(7) for $d_Y d_Y^c \beta(t, T)$. First, using Lemmas 6 and 7, we get

$$\lim_{T \to \infty} d_Y d_Y^c \beta(t, T)^* =$$

$$i_t^* b(z)^* - \sum_{q \geq 0} (-1)^q \mathrm{ch}(R^q f_* E, L^2) - t \log(t) i_t^* \left( \frac{\partial b}{\partial r} \right)^*.$$

Moreover, the estimate $b(z)^{[2]} = O(\frac{1}{|z|^3})$ shows the existence of

$$\beta(t) = \lim_{T \to \infty} \beta(t, T) = \int_{|z| \geq t} b(z) \, \log|z|^2,$$

and we may differentiate with respect to the $Y$-coordinates to get:

(15) $\quad d_Y d_Y^c \beta(t)^* = i_t^* b(z)^* - \sum_{q \geq 0} (-1)^q \mathrm{ch}(R^q f_* E, L^2) - t \log(t) i_t^* \left( \frac{\partial b}{\partial r} \right)^*.$

We now study the behavior as $t \to 0$ of both sides of (15). First recall that

(16) $$\beta(t) = \int_{|z| \geq t} b(z) \log|z|^2,$$

and, from (9), the component of $b(z)$ of degree 0 on $Y$ is

$$b(z)^{(0)} = \mathrm{tr}_s \exp(-|z|^2 \Delta - dz \, \overline{\partial} - d\overline{z} \, \overline{\partial}^*).$$

Using Duhamel's formula (Chapter V, (15)) and the fact that $\mathrm{tr}_s$ vanishes on supercommutators, this is equal to

$$b(z)^{(0)} = \mathrm{tr}_s((1 - dz \, \overline{\partial}) \, \exp(-|z|^2 \Delta - d\overline{z} \, \overline{\partial}^*)).$$

Now $\Delta$ commutes with $\overline{\partial}$ and $\overline{\partial}^*$. So we can use again Duhamel's formula and the fact that $\mathrm{tr}_s$ vanishes on supercommutators to get

(17) $$b(z)^{(0)} = \mathrm{tr}_s((1 - dz\overline{\partial})(1 - d\overline{z}\overline{\partial}^*) \exp(-|z|^2 \Delta)).$$

Because $\Delta$ is a smooth family of generalized Laplacians on $X \xrightarrow{f} Y$ we deduce from this formula the existence of an asymptotic development, as $t \to 0$,

(18) $$b(t)^{(0)} = \sum_{i \geq -n} b_i^{(0)} t^i \quad .$$

From (16) and (18) we get an asymptotic development

$$\beta(t)^{(0)} = \sum_{i \geq -n} \beta_i t^i + \sum_{j \geq -n} \gamma_j t^j \log(t),$$

where $\beta_i$ and $\gamma_j$ are smooth functions on $Y$. In particular, we may consider

(19) $$\beta_0 =: \text{finite part of} \int_{|z| \geq t} b(z)^{(0)} \log|z|^2,$$

which plays the role of a first Bott–Chern class for the Dolbeault complex.

Now let us look at $i_t^* \left(\frac{\partial b}{\partial r}\right)^*$ as $t$ goes to zero. Recall from (14) that

$$\frac{\partial}{\partial r}(b(z))dr = -d_Y\, b(z).$$

Using polar coordinates $z = re^{i\theta}$ and (9), we get

(20)
$$b(z) = \mathrm{tr}_s\, \exp(-(\tilde{\nabla}^2 + r^2\Delta + r\tilde{\nabla}(\bar{\partial}) + r\tilde{\nabla}(\bar{\partial}^*) + dr(\bar{\partial} + \bar{\partial}^*) + i\, rd\theta(\bar{\partial} - \bar{\partial}^*))).$$

The component of $b(z)^{(1)}$ involving only $dr$ may be computed from (20) as in (17). It is equal to

(21)
$$\mathrm{tr}_s\, \exp(-(r^2\Delta + r\tilde{\nabla}(\bar{\partial}) + r\tilde{\nabla}(\bar{\partial})^* + dr(\bar{\partial} + \bar{\partial}^*)))^{(1)}$$
$$= r\, \mathrm{tr}_s((\bar{\partial} + \bar{\partial}^*)(\tilde{\nabla}(\bar{\partial} + \bar{\partial}^*)))\exp(-r^2\Delta))dr.$$

Using (21) and applying $d_Y$ to (20), we see that $i_t^* \left(\frac{\partial b}{\partial r}\right)^*$ has an asymptotic development $\sum_{i \geq -n} \alpha_i t^i$ as $t \to 0$.

Finally, recall from (8) that

$$\lim_{t \to 0} [i_t^* b(z)^*]^{(2)} = \left[\int_{X/Y} \mathrm{ch}(E,h)Td(f)\right]^{(2)}.$$

So, if we take the finite part of the component of degree two on $Y$ in (15), we get, by the uniqueness of asymptotic developments,

(22)
$$d_Y d_Y^c(\beta_0)^* = f_*(\mathrm{ch}(E,h)Td(f))^{(2)} - c_1(\lambda(E)_{L^2}).$$

**5.2**   What remains to be proved is the following

**Lemma 8**   $(d_Y d_Y^c \beta_0^{(0)})^* = -d_Y d_Y^c T(E)$, *where* $T(E)$ *is the analytic torsion.*

*Proof.*   From (17) we deduce that the coefficient of $rdrd\theta$ in $b(z)^{(0)}$ is equal to

$$-i\mathrm{tr}_s((\bar{\partial} + \bar{\partial}^*)(\bar{\partial} - \bar{\partial}^*)\exp(-|z|^2\Delta)).$$

Therefore

$$\beta(t)^{(0)} = -4\pi i \int_t^\infty \mathrm{tr}_s((\bar{\partial} + \bar{\partial}^*)(\bar{\partial} - \bar{\partial}^*)\exp(-r^2\Delta))r\log r\, dr.$$

Let $N$ be the "number operator" acting by multiplication by $q$ on $D^q$. Clearly $[N, \bar{\partial}] = \bar{\partial}$ and $[N, \bar{\partial}^*] = -\bar{\partial}^*$. So

$$(\bar{\partial} + \bar{\partial}^*)(\bar{\partial} - \bar{\partial}^*) = (\bar{\partial} + \bar{\partial}^*)[N, \bar{\partial} + \bar{\partial}^*] = (\bar{\partial} + \bar{\partial}^*)N(\bar{\partial} + \bar{\partial}^*) - N\Delta.$$

Furthermore, since $\bar{\partial} + \bar{\partial}^*$ commutes with $\Delta = (\bar{\partial} + \bar{\partial}^*)^2$,

$$\mathrm{tr}_s(((\bar{\partial} + \bar{\partial}^*)N(\bar{\partial} + \bar{\partial}^*)\exp(-r^2\Delta))$$
$$= -\mathrm{tr}_s(N(\bar{\partial} + \bar{\partial}^*)\exp(-r^2\Delta)(\bar{\partial} + \bar{\partial}^*))$$
$$= -\mathrm{tr}_s(N\Delta\exp(-r^2\Delta)).$$

So

$$\beta(t)^{(0)} = 8\pi i \int_t^\infty \text{tr}_s(N\Delta \exp(-r^2\Delta))r \log r \ dr.$$

The change of variable $u = r^2$ gives

$$\beta(t)^{(0)} = 2\pi i \int_{t^2}^\infty \text{tr}_s(N\Delta \exp(-u\Delta)) \log u \ du.$$

Let $\theta(u) = tr'_s(N \exp(-u\Delta))$ be the supertrace of the restriction of $N \exp(-u\Delta)$ to the orthogonal complement to $\text{Ker}(\Delta)$. We have

$$\frac{\partial}{\partial u}\theta(u) = -\text{tr}_s(N\Delta \exp(-u\Delta)).$$

Integrating by parts, we find:

$$(23) \qquad \beta(t)^{(0)} = 2\pi i\theta(t^2) \log(t) + 2\pi i \int_{t^2}^\infty \theta(u) \frac{du}{u}.$$

We know that $\theta(u)$ as an asymptotic development $\sum_{i \geq -n} \theta_i u^i$ as $u \to 0$, so

$$\beta_0^{(0)} = \text{finite part of } 2\pi i \int_{t^2}^\infty \theta(u)\frac{du}{u}.$$

By definition of $N$, we also have

$$\sum_{q \geq 0}(-1)^q q\zeta_q(s) = \frac{1}{\Gamma(s)} \int_0^\infty \theta(u)u^s \frac{du}{u}.$$

Call this function $\zeta(s)$. By Theorem V.1, with $\epsilon = t^2$, we get

$$(24) \qquad \frac{1}{2\pi i}\beta_0^{(0)} = \zeta'(0) - \gamma\zeta(0).$$

As $\zeta'(0) = -T(E)$ the conclusion of Lemma 8 will follow from

**Lemma 9**   *On $Y$ we have $d_Y d_Y^c \zeta(0) = 0$.*

*Proof.* (This proof was explained to us by J.-M. Bismut.) Let $a > 0$ be a positive real number and let us multiply the metric $h_f$ on $Tf$ by $a$. By definition (see §V.2.2), the $L^2$-metric on $A^{0,q}(X_y, E), y \in Y(\mathbb{C})$, is multiplied by $a^{d-q}$ for all $q \geq 0$. It follows that $\bar{\partial}^*$, hence $\Delta$, is multiplied by $a$. From Lemma V.2, $\zeta(0)$ is unchanged and $\zeta'(0)$ gets replaced by $\zeta'(0) + \log(a)\zeta(0)$. Therefore, by (24), $\frac{1}{2\pi i}\beta_0^{(0)}$ becomes $\frac{1}{2\pi i}\beta_0^{(0)} + \log(a)\zeta(0)$.

On the other hand, let us look at the right-hand side of equation (22). We claim that it does not change when multiplying $h_f$ by $a$. Indeed, the hermitian holomorphic connection of $(Tf, h_f)$, hence the form $Td(Tf, h_f)$, remains unchanged. Furthermore, the $L^2$-metric on $\lambda(E)$ is multiplied by a constant, hence $c_1(\lambda(E)_{L^2})$ is also unchanged.

From these facts and (22) applied to $h_f$ and $ah_f$ we conclude that $d_Y d_Y^c \zeta(0) = 0$.                                                                                    $\square$

**5.4**     We can now conclude the proof of:

**Theorem VI.4**     *Assume that $f$ is locally Kähler, i.e., for every $y \in Y$, there is a neighborhood $U$ of $y$ such that $f^{-1}(U)$ has a Kähler metric. Then*

$$c_1(\lambda(E)_Q) = f_*(\mathrm{ch}(E,h)Td(Tf,h_f))^{(2)},$$

*where $Td(Tf,h_f)$ is the Todd form of $(Tf,h_f)$ and $(.)^{(2)}$ denotes the component of degree 2.*

*Proof.* Recall that we can assume that $R^q f_* E$ is locally free on $Y$. By combining Lemma 8 and the equality (22), we then get

$$-dd^c T(E) = f_*(\mathrm{ch}(E,h)Td(f))^{(2)} - c_1(\lambda(E)_{L^2}).$$

But $h_Q = h_{L^2} \exp(T(E))$ so

$$c_1(\lambda(E)_Q) = c_1(\lambda(E)_{L^2}) - dd^c T(E).$$

The conclusion follows.     □

### 5.5 Remark

It would be interesting to give an *axiomatic* definition of Quillen's metric, analogous to the axioms of Theorem IV.2 for the Bott–Chern character forms. We know the curvature of Quillen's metric, but how can it be normalized? Finding such an axiomatic description might lead to a simpler proof of the arithmetic Riemann–Roch–Grothendieck theorem (Theorem VIII.1′), by checking the axioms for both sides of the equality; §V.1.3.3 might be of some use for this project.

# VIII

## The Arithmetic
## Riemann–Roch–Grothendieck
## Theorem

In this final chapter we shall give an arithmetic Riemann–Roch–Grothendieck theorem (Theorem 1) for the determinant of the cohomology, and deduce from it an estimate for the smallest size of non-zero sections of powers of ample line bundles on arithmetic varieties, a result we announced in the Introduction.

The arithmetic Riemann–Roch–Grothendieck theorem (Theorem 1) combines the, somewhat classical, algebraic Riemann–Roch–Grothendieck theorem for higher Chow groups with the computation we made in the previous chapter of the curvature of the Quillen metric on the determinant of the cohomology. There exists a more precise theorem, Theorem 1', but its proof is very long and will not be given here. It is not needed for the application.

To get bounded sections of powers of ample line bundles (Theorem 2) we combine Theorem 1 with Minkowski's theorem, a lemma of Gromov, and a result of Bismut–Vasserot about the asymptotic behavior of the analytic torsion of powers of a given line bundle, the proof of which is only sketched.

### 1. The arithmetic Riemann–Roch–Grothendieck theorem

**1.1**    Let $f : X \to Y$ be a projective and flat map between arithmetic varieties, i.e. regular, projective flat schemes over $\operatorname{Spec} \mathbb{Z}$. We assume that the restriction of $f$ to the generic fiber $X_{\mathbb{Q}}$ is smooth. Let $\overline{E} =$

$(E, h)$ be an hermitian vector bundle over $X$. Finally, choose a metric on the relative tangent space $Tf|_{X(\mathbb{C})}$ invariant under complex conjugation and such that the restriction to $f^{-1}(y)$ is Kähler for every $y \in Y(\mathbb{C})$. We may consider, as in Chapter VI, the determinant of cohomology, i.e. the line bundle $\lambda(\overline{E})_Q$ on $Y$, equipped with its Quillen metric. Our goal is to give a formula for its first Chern class $\hat{c}_1(\lambda(\overline{E})_Q) \in \widehat{CH}^1(Y)$.

To the hermitian bundle $\overline{E}$ is associated its arithmetic Chern character

$$\widehat{\mathrm{ch}}(E, h) \in \widehat{CH}(X)_\mathbb{Q} = \bigoplus_{p \geq 0} \widehat{CH}^p(X)_\mathbb{Q},$$

defined in §IV.4. We need also a Todd genus for the map $f$. To define it, let us consider a factorisation of $f$:

$$
\begin{array}{ccc}
X & \xrightarrow{\ i\ } & \mathbb{P}(E)_Y \\
\ _f\searrow & & \downarrow p \\
& Y\,, &
\end{array}
$$

where $i$ is a closed embedding, and $p$ a projective bundle over $Y$. Let $N$ be the normal bundle of $X$ in $\mathbb{P}(E)_Y$, and $i^*Tp$ the restriction to $X$ of the relative tangent bundle to the projection $p$. On $X(\mathbb{C})$ we have an exact sequence of holomorphic bundles

$$\mathcal{E} : 0 \longrightarrow Tf_\mathbb{C} \longrightarrow i^*Tp_\mathbb{C} \longrightarrow N_\mathbb{C} \longrightarrow 0.$$

Let us choose hermitian metrics on $N$ and $i^*Tp$. As was noticed in §IV.3.3 (see also [GS3]), by the same construction as in Theorem IV.2, one gets a secondary characteristic class $\widetilde{Td}(\mathcal{E}) \in \tilde{A}(X)$ such that

(1)  $$dd^c(\widetilde{Td}(\mathcal{E})) = Td(\overline{Tf})Td(\overline{N}) - Td(i^*\overline{Tp}).$$

Moreover we have the arithmetic Todd classes $\widehat{Td}(i^*\overline{Tp}) \in \widehat{CH}(X)_\mathbb{Q}$ and $\widehat{Td}(\overline{N}) \in \widehat{CH}(X)_\mathbb{Q}$ defined in §IV.4.6.

Now we define

(2)  $$\widehat{Td}(f) = \widehat{Td}(i^*\overline{Tp})\widehat{Td}(\overline{N})^{-1} + a(\widetilde{Td}(\mathcal{E})Td(\overline{N})^{-1}) \in \widehat{CH}(X)_\mathbb{Q};$$

recall that the map $a : \tilde{A}(X) \to \widehat{CH}(X)$ was defined in Chapter III as sending $\eta$ to the class of $(0, \eta)$. If $f$ is smooth, $\widehat{Td}(f)$ is the Todd class of the relative tangent bundle to $f$. It can be shown in general that it is independent of the choice of $i, p$ and the metrics on $N$ and $i^*Tp$; see [GS8].

We may now consider the element

(3)  $$\delta(\overline{E}) = \hat{c}_1(\lambda(\overline{E})_Q) - f_*(\widehat{\mathrm{ch}}(E, h)\widehat{Td}(f))^{(1)}$$

in $\widehat{CH}^1(X)_\mathbb{Q}$, where $\alpha^{(1)}$ denotes the component of degree one of $\alpha \in \widehat{CH}(X)_\mathbb{Q}$.

**Theorem 1**   *There is a morphism*
$$\delta : K_0(X_{\mathbb{Q}}) \to H^{0,0}(Y)/\rho(CH^{1,0}(Y))_{\mathbb{Q}}$$
*such that* $a(\delta([E])) = \delta(\overline{E})$.

*Proof.* We first show that $\delta(\overline{E})$ lies in $H^{0,0}(Y)/\rho(CH^{(1,0)}(Y))_{\mathbb{Q}}$. Recall from Theorem III.1 that there is an exact sequence
$$CH^{1,0}(Y) \xrightarrow{\rho} H^{0,0}(Y) \xrightarrow{a} \widehat{CH}^1(Y) \xrightarrow{(z,\omega)} CH^1(Y) \oplus A^{1,1}(Y).$$
We have to show that $(z,\omega)(\delta(\overline{E})) = 0$. The cycle component is
$$z(\delta(\overline{E})) = c_1(\lambda(E)) = f_*(\mathrm{ch}(E)Td(f))^{(1)} \in CH^1(Y)_{\mathbb{Q}}.$$
It vanishes because of the Riemann–Roch–Grothendieck theorem for (classical) Chow groups; see [GBI], [S1].

Let us now compute the form $\omega(\delta(\overline{E}))$. From (1) and (2) we deduce that
$$\omega(\widehat{Td}(f)) = Td(i^*\overline{Tp})Td(\overline{N})^{-1} + dd^c(\tilde{T}d(\mathcal{E})Td(\overline{N})^{-1}) = Td(\overline{Tf_{\mathbb{C}}})$$
in $A(X)$. Therefore
$$\omega(f_*(\widehat{\mathrm{ch}}(\overline{E})\widehat{Td}(f)))^{(2)} = f_*(\mathrm{ch}(E,h)\omega(\widehat{Td}(f)))^{(2)}$$
$$= f_*(\mathrm{ch}(E,h)Td(Tf,h_f))^{(2)}$$
$$= c_1(\lambda(E)_Q) \quad \text{by Theorem V.4.}$$
In other words $\omega(\delta(\overline{E})) = 0$ and $\delta(\overline{E})$ is the image by $a$ of a class in $H^{0,0}(Y)/(\mathrm{im}\,\rho)_{\mathbb{Q}}$.

Our second step is to show that $\delta(\overline{E})$ depends only on the class of $E$ in $K_0(X)$. Let
$$\mathcal{E} : 0 \longrightarrow S \longrightarrow E \longrightarrow Q \longrightarrow 0$$
be an exact sequence of bundles on $X$, endowed with metrics $h', h, h''$. We want to show that $\delta(\overline{E}) = \delta(\overline{S}) + \delta(\overline{Q})$. By the Corollary VI.1, we have
$$\hat{c}_1(\lambda(\overline{E})_Q) = \hat{c}_1(\lambda(\overline{S})_Q) + \hat{c}_1(\lambda(\overline{Q})_Q) - a\left(\int_{X/Y} \tilde{\mathrm{ch}}(\mathcal{E})Td(\overline{Tf_{\mathbb{C}}})\right)^{(0)}.$$
Furthermore, by Proposition IV.1,
$$\widehat{\mathrm{ch}}(E,h) = \widehat{\mathrm{ch}}(S,h') + \widehat{\mathrm{ch}}(Q,h'') - a(\tilde{\mathrm{ch}}(\mathcal{E})),$$
therefore
$$f_*(\widehat{\mathrm{ch}}(\overline{E})\widehat{Td}(f))$$
$$= f_*(\widehat{\mathrm{ch}}(\overline{S})\widehat{Td}(f)) + f_*(\widehat{\mathrm{ch}}(\overline{Q})\widehat{Td}(f)) + f_*(a(\tilde{\mathrm{ch}}(\mathcal{E}))\widehat{Td}(f)).$$
Finally, using Chapter III, (3), we get
$$f_*(a(\tilde{\mathrm{ch}}(\mathcal{E}))\widehat{Td}(f)) = a\left(\int_{X/Y} \tilde{\mathrm{ch}}(\mathcal{E})Td(\overline{Tf_{\mathbb{C}}})\right).$$

We conclude that $\delta(\overline{E}) = \delta(\overline{S}) + \delta(\overline{Q})$. In other words, $\delta(\overline{E})$ is independent of the metric on $E$, and there is a morphism

$$\delta : K_0(X) \longrightarrow H^{0,0}(Y)/(\operatorname{im}\rho)_{\mathbb{Q}}$$

such that $a(\delta([E])) = \delta(\overline{E})$.

Finally, let us prove that $\delta$ factors via $K_0(X_{\mathbb{Q}})$. We have

$$\ker(K_0(X) \longrightarrow K_0(X_{\mathbb{Q}})) = \operatorname{im}(K_0(X)_{\mathrm{fin}} \longrightarrow K_0(X))$$

where $K_0(X)_{\mathrm{fin}} = \bigoplus_p K_0^{X_p}(X)$ is the Grothendieck group of $X$ with support in finite fibers. There is a commutative diagram:

$$
\begin{array}{ccc}
K_0(X)_{\mathrm{fin}} & \longrightarrow & \widehat{K}_0(X) \\
{\scriptstyle \mathrm{ch}} \downarrow & & \downarrow {\scriptstyle \widehat{\mathrm{ch}}} \\
CH(X)_{\mathrm{fin}} & \xrightarrow{\ \varphi\ } & \widehat{CH}(X)
\end{array}
$$

where $\varphi$ maps the cycle $Z$ to the class of $(Z,0)$. If $x \in K_0(X)_{\mathrm{fin}}$ the equality

$$\delta(x) = \varphi(c_1(f_*(x)) - f_*(\mathrm{ch}(x)Td(f))^{(1)}) = 0$$

follows from the Riemann–Roch–Grothendieck theorem with supports [S1]. So $\delta$ factors via $K_0(X_{\mathbb{Q}})$. $\qquad\square$

**1.2**   It follows from Theorem VI.5 that $\delta([E])$ is independent of the choice of the metric on $Tf_{\mathbb{C}}$. In fact there is a precise formula for $\delta$. To state it, we let $\zeta(s)$ be the Riemann zeta function and $\zeta'(s)$ its derivative. Given an holomorphic vector bundle $E$ on a complex manifold $X$, we may define as follows a characteristic class $R(E) \in H^*(X,\mathbb{R})$ in the real cohomology of $X$. The class $R$ commutes with pull-back $(R(f^*(E)) = f^*(R(E)))$, it is additive on exact sequences, and its value on a line bundle $L$ with first Chern class $c_1(L) \in H^2(X,\mathbb{R})$ is

$$R(L) = \sum_{m \text{ odd} \geq 1} (2\zeta'(-m) + \zeta(-m)(1 + \frac{1}{2} + \cdots + \frac{1}{m}))\frac{c_1(L)^m}{m!} \in H^*(X).$$

In the situation of §1.1, let $\mathrm{ch}(E_{\mathbb{C}}) \in H^*(X,\mathbb{R})$ be the usual Chern character of the holomorphic bundle $E_{\mathbb{C}}$ induced by $E$, and $Td(Tf_{\mathbb{C}}) \in H^*(X,\mathbb{R})$ the Todd class of the relative tangent space. Recall that $H^{p,p}(Y) \subset \widetilde{A^{p,p}}(Y)$ is mapped by $a$ into $\widehat{CH}(Y)_{\mathbb{Q}}$.

**Theorem 1'**   [GS7] *The following equality holds:*

$$\delta(\overline{E}) = a(f_*(-\mathrm{ch}(E_{\mathbb{C}})Td(Tf_{\mathbb{C}})R(Tf_{\mathbb{C}}))^{(0)}).$$

**1.3**    The proof of Theorem 1′ is much harder than the proof of Theorem 1. It was obtained in [GS7] and [GS8] by combining the work of several people ([B2], [B3], [BL], [BGS2], [BGS3], [GS5]). It can be extended to the case where $X$ and $Y$ are only quasi-projective, $Y$ is regular, and the restriction of $f$ to the generic fiber $X_{\mathbb{Q}}$ is smooth (but $X$ need not be regular); [GS7], [GS8] Theorem 7.

In [F4], Faltings extended this result to higher degrees, when $X$ and $Y$ are regular. Namely, under the assumptions of Theorem 1′, using higher degree analogs of the analytic torsion, one can define a direct image group morphism for virtual hermitian vector bundles (in the sense of §IV.4.8)

$$f_* : \widehat{K}_0(X) \longrightarrow \widehat{K}_0(Y)$$

(see [GS9], [GS5], [F4], [BKo]; the precise comparison of the different definitions has still to be made). The determinant of $f_*(\overline{E})$ is the class of $\lambda(\overline{E})_{\mathbb{Q}}$ in $\widehat{\mathrm{Pic}}(Y)$. Theorem 6.1 in [F4] gives a formula for $\widehat{\mathrm{ch}} \circ f_*$ extending Theorem 1′.

**1.4**    When $Y = \mathrm{Spec}\,\mathbb{Z}$, we have, by Chapter III, (7),

$$\widehat{CH}^1(Y) = \widehat{\mathrm{Pic}}(\mathbb{Z}) = \mathbb{R},$$

and $\delta(\overline{E})$ is a real number. Theorems 1 and 1′ appear then as an arithmetic analog of the Riemann-Roch-Hirzebruch theorem [Hz]. Indeed, the following lemma shows that $\hat{c}_1(\lambda(\overline{E})_{\mathbb{Q}})$ is an Euler characteristic. For each $q \geq 0$, $H^q(X, E)$ is a finitely generated abelian group. Let $H^q(X, E)_{\mathrm{Tors}}$ be its torsion subgroup and $\mathrm{Vol}_{L^2}(H^q(X, E))$ the volume for the $L^2$-norm of the quotient $H^q(X(\mathbb{C}), E_{\mathbb{C}})^+ / H^q(X, E)$, where $(.)^+$ denotes the real subspace invariant under the complex conjugation. Let $\zeta_q(s)$ be the zeta function of $\Delta^q = \overline{\partial}\,\overline{\partial}^* + \overline{\partial}^*\overline{\partial}$ in degree $(0, q)$.

**Lemma 1**    *In* $\widehat{CH}^1(\mathbb{Z}) = \mathbb{R}$ *we have*

$\hat{c}_1(\lambda(\overline{E})_{\mathbb{Q}})$

$$= \sum_{q\geq 0}(-1)^q \left( \log \sharp H^q(X, E)_{\mathrm{Tors}} - \log\ \mathrm{Vol}_{L^2}\left(H^q(X, E)\right) + \frac{1}{2}q\zeta_q'(0) \right).$$

*Proof.* Since $\mathrm{Spec}\,\mathbb{Z}$ is affine, we have

$$\lambda(E) = \bigotimes_{q\geq 0} \det R^q f_*(E)^{(-1)^q} = \bigotimes_{q\geq 0} \det H^q(E)^{(-1)^q}.$$

By the definition of Quillen metric (Chapter VI), the lemma will follow from the following general fact. Let $\overline{M}$ be a finitely generated $\mathbb{Z}$-module $M$, equipped with an hermitian scalar product on $M \otimes_{\mathbb{Z}} \mathbb{C}$, invariant under conjugation. Its determinant $\det \overline{M}$ has a degree $\widehat{\deg} \det \overline{M} \in \mathbf{R}$. We claim that

$$\widehat{\deg} \det \overline{M} = \log \sharp M_{\text{Tors}} - \log \text{Vol}(\overline{M}).$$

Since this equality is compatible with direct sums, it is enough to check it in two cases. First, assume that $M$ is torsion free. Then the line $\det M = \Lambda^{\max} M$ is generated by $s_1 \wedge \ldots \wedge s_r$, where $s_1, \ldots, s_r$ is any basis of $M$. Then

$$\widehat{\deg} \det \overline{M} = -\log \|s_1, \wedge \ldots \wedge s_r\| = -\log \text{Vol} \overline{M}.$$

On the other hand, if $M = \mathbf{Z}/n$ we have a free resolution

$$0 \longrightarrow \mathbf{Z} \xrightarrow{\;\varphi\;} \mathbf{Z} \longrightarrow \mathbf{Z}/n \longrightarrow 0 ,$$
$$s_1 \longmapsto n s_0$$

where $s_0$ and $s_1$ are the generators. The line $\det M$ is generated by $s_0 \otimes s_1^{-1}$ and

$$\widehat{\deg} \det \overline{M} = -\log(\|s_0\|/\|\varphi(s_1)\|) = \log n.$$

This proves Lemma 1.

**1.5**

**Questions**    Does there exist an "arithmetic Lefschetz theorem" generalizing the previous result to the case of group actions?

Can one extend to varieties over number fields Grothendieck's cohomological description of zeta functions of varieties over finite fields [Gro] by means of a Lefschetz theorem?

## 2. Arithmetic ampleness

**2.1**    Recall from [H] [GH] that for an algebraic variety $X$ over $\mathbf{C}$, we have the following equivalent definitions for a line bundle $L$ to be *ample*:

(1)    For every coherent sheaf $\mathcal{E}$ over $X$ there is $n_0$ such that, for every $n > n_0$, $\mathcal{E} \otimes L^{\otimes n}$ is generated by global sections;

(2)    For every coherent sheaf $\mathcal{E}$ over $X$ there is $n_0$ such that, for every $n > n_0$ and $k > 0$,

$$H^k(X, \mathcal{E} \otimes L^{\otimes n}) = 0;$$

(3)    There is $n_0$ and some embedding $\varphi : X \hookrightarrow \mathbf{P}^N$ such that $L^{\otimes n_0} = \varphi^*(\mathcal{O}(1))$;

(4)    There is a metric $h$ on $L$ such that $c_1(L, h) > 0$, i.e., for every tangent vector $v \in T_x(X)$,

$$c_1(L, h)(v, \overline{v}) > 0.$$

**2.2**     Let $X$ be a regular, projective, flat scheme over $\operatorname{Spec} \mathbf{Z}$. Let $\overline{L}$ be a hermitian line bundle over $X$. We assume that $L$ is ample over $X$, i.e. Property (1) in §2.1 holds on $X$, and that $c_1(L, h) > 0$ over $X(\mathbf{C})$. Let $\overline{E}$ be an arbitrary hermitian vector bundle over $X$, and let $r = \operatorname{rank} E$. We shall look for a nontrivial global section $s \in \Gamma(E \otimes L^{\otimes n})$ with a uniform bound on $\|s(x)\|$ , $x \in X(\mathbf{C})$.

Let $f : X \longrightarrow \operatorname{Spec} \mathbf{Z}$ be the map of definition. We have two relevant self-intersection numbers: the geometric one

$$L^d = f_*(c_1(L)^d) \in H^0(\operatorname{Spec} \mathbf{C}, \mathbf{Z}) = \mathbf{Z},$$

and the arithmetic one

$$\overline{L}^{d+1} = f_*(\hat{c}_1(\overline{L})^{d+1}) \in \widehat{CH}^1(\mathbf{Z}) = \mathbf{R}.$$

**Theorem 2**     *Let $n \geq 1$ be an integer and $\epsilon > 0$ a real number. Then, the logarithm of the number of sections $s \in H^0(X, E \otimes L^{\otimes n})$ such that, for every $x \in X(\mathbf{C})$,*

$$\|s(x)\|^2 < \exp(n(\epsilon - \overline{L}^{d+1}/(d+1)L^d))$$

*is at least equal to*

$$\epsilon r L^d \; n^{d+1}/2d! + O(n^d \log n).$$

*In particular, when $n$ is big enough there is a nontrivial section satisfying these inequalities.*

The proof will actually show that we need not assume that $L$ is ample over $X$, but only that, for all $p > 0$, $H^{2p}(X, E \otimes L^{\otimes n}) = 0$ for large $n$, and $c_1(\overline{L}) > 0$.

We will reduce the theorem to the following result, which is reminiscent of the Nakai–Moishezon criterion for ampleness; see [H], V, Theorem 1.10.

**Theorem 2'**     *Assume that $\overline{L}^{d+1} > 0$. Then the logarithm of the number of sections $s \in H^0(X, E \otimes L^{\otimes n})$ such that, for every $x \in X(\mathbf{C})$, $\|s(x)\| < 1$ is greater or equal to*

$$r\overline{L}^{d+1} \; n^{d+1}/(d+1)! + O(n^d \log n).$$

To see that Theorem 2 implies Theorem 2', take $\epsilon = \overline{L}^{d+1}/(d+1)L^d$.

To show that Theorem 2' implies Theorem 2, let $\overline{L} = (L, h)$. When $\lambda \in \mathbf{R}$, consider the metric $he^\lambda$, and write $\overline{L}_\lambda = (L, he^\lambda)$. We first notice that

$$(4) \qquad \overline{L}_\lambda^{d+1} = [\overline{L} \otimes (\mathcal{O}_X, e^\lambda)]^{d+1} = \overline{L}^{d+1} - \frac{1}{2}(d+1)\lambda L^d.$$

To see this, observe that

$$\hat{c}_1(\overline{L} \otimes (\mathcal{O}_X, e^\lambda)) = \hat{c}_1(\overline{L}) - a(\lambda),$$

where $\lambda \in H^0(X(\mathbb{C}), \mathbb{R})$ is sent to $a(\lambda)$ in $\widehat{CH}^1(X)$ as in Theorem III.1. From the rule $a(\eta)y = a(\eta\omega(y))$ of §III.2.3.1 we get $a(\lambda)a(\lambda) = a(\lambda)\omega(a(\lambda)) = 0$. Therefore

$$
\begin{aligned}
\hat{c}_1(\overline{L}_\lambda)^{d+1} &= \hat{c}_1(\overline{L})^{d+1} + (d+1)a(-\lambda) \cdot \hat{c}_1(\overline{L})^d \\
&= \hat{c}_1(\overline{L})^{d+1} + (d+1)a(-\lambda\, c_1(L)^d).
\end{aligned}
$$

(5)

Now, from §§III.4.2 and 4.3, we get:

$$
\widehat{\deg}(f_*a(\alpha)) = \frac{1}{2}\int_{X(\mathbb{C})} \alpha.
$$

So we get (4) by applying $f_*$ to (5).

Now choose $\lambda$ so that

$$
\overline{L}_\lambda^{d+1} = \overline{L}^{d+1} - \frac{d+1}{2}\lambda L^d = (d+1)L^d\epsilon > 0.
$$

From Theorem 2' applied to $\overline{L}_\lambda$ we get

$$
\log \#\{s \in H^0(X, E \otimes L^{\otimes n}); \forall x \in X(\mathbb{C})\ \|s(x)\| < e^{-n\lambda/2}\}
$$

$$
\geq (\epsilon\, \frac{r}{d!}\, \frac{L^d}{2})n^{d+1} + O(n^d \log n).
$$

$\square$

**2.3**    We shall now decribe the proof of Theorem 2'. Let

$$
\Lambda_n = H^0(X, E \otimes L^{\otimes n}).
$$

This is a lattice in the real subspace $H^0(X(\mathbb{C}), E \otimes L^{\otimes n})^+$ of $H^0(X(\mathbb{C}), E \otimes L^{\otimes n})$. On this vector space we may consider the $L^2$-norm, or the sup-norm. Then take the Haar measure for which the unit ball for the $L^2$-norm (resp. the sup-norm) has volume equal to the volume $V_n$ of the unit ball in the standard euclidean space whose dimension is the rank of $\Lambda_n$. Denote by $\mathrm{Vol}_{L^2}(\Lambda_n)$ (resp. $\mathrm{Vol}_{\sup}(\Lambda_n)$) the covolume of $\Lambda_n$ for this measure. By Minkowski's theorem, see [GL] II.7.2, Theorem 1, we know that:

(6)
$$
\log \#\{s \in H^0(X, E \otimes L^{\otimes n}); \|s\|_{\sup} \leq 1\}
$$
$$
\geq -\log \mathrm{Vol}_{\sup}(\Lambda_n) + \log V_n - \log 2 \cdot \mathrm{rank}(\Lambda_n).
$$

Now $\mathrm{rank}(\Lambda_n) = O(n^d)$ by the classical theory of Hilbert polynomials, see [H] I.7, and the vanishing of $H^k(X(\mathbb{C}), E \otimes L^{\otimes n})$, $k > 0$. The usual formula for the volume of the euclidean unit ball and Stirling's formula (Chapter V, (2)) imply that

(7)    $$\log V_n = O(n^d \log n).$$

Furthermore we shall prove the following inequalities, where $T(E \otimes L^{\otimes n})$ is the analytic torsion (Definition VI.3) of $E \otimes L^{\otimes n}$ for the chosen metrics.

(8) $-\log \mathrm{Vol}_{\sup}(\Lambda_n) \geq -\log \mathrm{Vol}_{L^2}(\Lambda_n) + O(n^d \log n)$

(9)
$$\geq -\log \mathrm{Vol}_{L^2}(\Lambda_n) - T(E \otimes L^{\otimes n})/2$$
$$+ O(n^d \log n)$$

(10)
$$= \widehat{c}_1(\lambda(E \otimes L^{\otimes n})_{\mathbb{Q}})$$
$$- \sum_{q \geq 1} (-1)^q \log \sharp H^q(X, E \otimes L^{\otimes n}) + O(n^d \log n)$$

(11)
$$\geq \left( \frac{r\overline{L}^{d+1}}{(d+1)!} \right) n^{d+1} + O(n^d \log n) + \delta(\overline{E} \otimes \overline{L}^n).$$

Recall now Corollary I.2 stating that $[E \otimes L^{\otimes n}] = O(n^d)$ in $K_0(X_{\mathbb{Q}})_{\mathbb{Q}}$. Since we know from Theorem 1 that $\delta$ is induced by a morphism on $K_0(X_{\mathbb{Q}})$ we get that $\delta(\overline{E} \otimes \overline{L}^n) = O(n^d)$. Combining this with (6), (7), (8), (9), (10) and (11), we get Theorem 2′.

To prove the inequality (11), we use the properties of the Chern character (Theorem IV.3) and compute
$$f_*(\widehat{\mathrm{ch}}(\overline{E} \otimes \overline{L}^n)\widehat{Td}(f)) = f_*(\widehat{\mathrm{ch}}(\overline{E}) \exp(n\widehat{c}_1(\overline{L}))\widehat{Td}(f)^{(d+1)}).$$
The leading term in $n$ is $(r\overline{L}^{d+1}/(d+1)!)n^{d+1}$, since $\widehat{Td}^0(f) = 1$ and $\widehat{\mathrm{ch}}^0(\overline{E}) = r$. From the definition of $\delta$ and the vanishing of $H^q$ for each (even) $q > 0$ and large $n$ we get (11).

To prove the equality (10), we apply Lemma 1 to the hermitian bundle $\lambda(E \otimes L^{\otimes n})_{\mathbb{Q}}$ over $\mathrm{Spec}\,\mathbb{Z}$.

**2.4**    To prove the inequality (9) we need to estimate the analytic torsion of $E \otimes L^{\otimes n}$ as $n$ goes to infinity. This was done by Bismut and Vasserot.

**Theorem 3** [BVa]    *Let $X$ be a compact Kähler manifold, $\overline{E}$ a holomorphic hermitian vector bundle and $\overline{L}$ a holomorphic hermitian line bundle on $X$. Assume that $c_1(\overline{L})$ is positive. Then*
$$T(E \otimes L^{\otimes n}) = O(n^d \log n).$$

We shall not give a detailed proof of this result but we indicate its main steps. Near every point $x \in X$ consider geodesic coordinates $y_\alpha$ and trivializations of $TX$, $E$ and $L$ by parallel transport. For every $q > 0$ the Laplace operator $\Delta^q$ can be written explicitly in terms of these coordinates ([BVa], (21)). One makes the (local) change of variable $\delta_n : y_\alpha \longrightarrow \sqrt{n}\, y_\alpha$ in this expression. It turns out that $\frac{1}{n}\delta_n(\Delta^q_{E \otimes L^{\otimes n}})$ has a limit as $n$ goes to infinity: call it $\mathcal{L}_x$. This operator $\mathcal{L}_x$ is elliptic, and the convergence holds in all $C^k$ norms. Therefore the heat kernel of $\frac{1}{n}\delta_n(\Delta_{E\otimes L^{\otimes n}}^q)$ converges to the heat kernel of $\mathcal{L}_x$, its asymptotic development as $t \to 0$ is uniform in $n$, and its coefficients $a_{i,n}$ converge to those of $\mathcal{L}_x$; this follows from the construction of the heat kernel as in §V.3. But the heat kernel of $\frac{1}{n}\delta_n(\Delta_{E\otimes L^{\otimes n}}^q)$ has the same trace as $\exp(-(t/n)\Delta_{E\otimes L^{\otimes n}}^q)/(rn^d)$, which is globally defined. Let

$$\Theta_n^q(t) = \mathrm{tr}(\exp(-(t/n)\Delta_{E\otimes L^{\otimes n}}^q)/(rn^d))$$

be this trace.

Using the fact that $c_1(\overline{L})$ is positive, Bismut and Vasserot prove that the smallest eigenvalue of $\Delta_{E\otimes L^{\otimes n}}^q$ is bounded below by $cn - c'$, where $c$ is a positive constant; this step is similar to the classical proof of the Kodaira Vanishing Theorem. It follows that there is a positive constant $c''$ such that, when $t$ goes to infinity, $\Theta_n^q(t) < \exp(-c''t)$. Therefore the sum

$$(12) \qquad \int_1^\infty \Theta_n^q(t)\frac{dt}{t} + \int_0^1 \left(\Theta_n^q(t) - \sum_{i\geq -d}\frac{a_{i,n}}{t^i}\right)\frac{dt}{t} + \sum_{i\geq -d}\frac{a_{i,n}}{i} + \gamma a_{0,n}$$

has a finite limit as $n \to \infty$. From (12) and the formula (11) in Chapter V for $\zeta'_{\Delta_{E\otimes L^{\otimes n}}^q}(0)$, we conclude that $T(E\otimes L^{\otimes n})/(n^d \log n)$ has a finite limit as $n \to \infty$, and Theorem 3 follows. $\qquad\square$

In fact a formula can be given for the first two main terms of $T(E\otimes L^{\otimes n})$ as $n \to \infty$, see [BVa]. This leads to a more precise version of Theorem 2′, see [GS8] Theorem 8.

**2.5**    Finally, to prove (8), we use a result of Gromov to compare the sup-norm with the $L^2$-norm on the sections of $E\otimes L^{\otimes n}$.

**Lemma 2** (Gromov)    *Let $X$ be a compact complex manifold, $\overline{E}$ a holomorphic hermitian vector bundle and $\overline{L}$ a holomorphic hermitian line bundle on $X$. Then, for every $n \geq 1$, there is a constant $c$ such that any section $s \in H^0(X, E\otimes L^{\otimes n})$ satisfies*

$$\|s\|_{\sup} \leq cn^d\|s\|_{L^2}.$$

This lemma implies that any section $s \in \Lambda^{\max}H^0(X, E\otimes L^{\otimes n})$ satisfies

$$\|s\|_{\sup} \leq (cn^d)^{\mathrm{rank}\Lambda_n}\|s\|_{L^2}$$

and, since $\mathrm{rank}\Lambda_n = O(n^d)$,

$$\log\|s\|_{\sup} \leq \log\|s\|_{L^2} + O(n^d \log n),$$

and (8) follows.

To prove Lemma 2 we choose a system of neighborhoods $U_x \subset X$ of every point $x \in X$ (varying continuously with $x$), with radius bounded below and biholomorphic to the unit ball $B_1$ in $\mathbb{C}^n$, and a trivialization of $E$ and $L$ on each $U_x$. If $x_0$ is the point where $\|s(x)\|$ is maximum, and if $\mu$ is the measure on $X$ then, writing $s = (f_1, \ldots, f_r)$ in $U_{x_0}$ we get

$$\int_X \|s(x)\|^2 \mu > \int_{B_1} |f(x)|^2 h(x) p(x)^n dx,$$

where $h(x)$ is a factor coming from the measure and the metric on $E$, $p(x)$ comes from the metric on $L$, and

$$|f(x)|^2 = \sum_{i \geq 0} |f_i(x)|^2$$

is subharmonic. But $h(x)$ is uniformly bounded from below, and $p(x)$ may be bounded from below by $p(x_0) - c_1 r$, where $r = |x - x_0|$ and $c_1$ is a uniform constant; we may need to take smaller neighborhoods for this. Therefore we get

$$\int_X \|s(x)\|^2 \mu$$

$$\geq c_2 \|f(0)\|^2 \int_{B_1} (p(x_0) - c_1 r)^n dx \geq c_3 \, n^{-d} |f(0)|^2 c_3 (p(x_0))^n$$

$$\geq c_4 \, n^{-2d} \|s\|_{\text{sup}}^2.$$

$\square$

## 3. Remarks

**3.1**    Theorem 2 was used by Vojta [V2] in his proof of the Mordell conjecture; notice however that, in the case considered in [V2], $c_1(\overline{L})$ is not necessarily positive on the whole of $X$. In the work of Faltings on abelian varieties [F3] and in Bombieri's version of Vojta's proof [Bo], it has been replaced by much more direct arguments.

**3.2**    The statement in Theorem 2′ asserts that $\overline{E} \otimes \overline{L}^{\otimes n}$ is *effective*, in an arithmetic sense, when $n$ is big enough. For $\overline{L}$ to be arithmetically *ample* would require that the sections of sup-norm less than one generate $E \otimes L^{\otimes n}$. Results of this kind have been obtained by Kim [Ki] and Zhang [Z] for arithmetic surfaces and by Zhang [Z2] in higher dimensions.

**3.3**    Theorem 2′ can be generalized to the case where $X$ is not necessarily regular: see [GS8] Theorem 8 when $X_{\mathbb{Q}}$ is smooth and [Z2] in general.

In this result, one would also like to be able to replace $E$ by a coherent sheaf. For instance, one can look for small sections of $L^{\otimes n}$ with high order of vanishing at a given finite set $S$ of points. This problem is well-known in the context of diophantine approximation, where it leads to the construction of *auxiliary polynomials*. Assume there are more small sections of $L^{\otimes n}$ than the number of values that their derivatives can take at $S$. By taking the difference of two such sections one gets a nontrivial small section of $L^{\otimes n}$ whose derivatives vanish at $S$; this argument is known as *Dirichlet's box principle*. The reader is referred to [Lf] for a precise result combining this principle with Theorem 2, thus generalizing the classical *Siegel's Lemma*. When $E$ is an arbitrary coherent sheaf, no general result exists yet.

# References

[A1] S.J. Arakelov: Intersection theory of divisors on an arithmetic surface, *Math. USSR Izvestija* **8** (1974), 1167-1180.

[A2] S.J. Arakelov: Theory of intersections on an arithmetic surface, *Proceedings International Congress of Mathematicians, Vancouver*, Vol. 1 (1975), 405-408.

[AT] M. Atiyah and D.O. Tall: Group representations, $\lambda$-rings and the $J$-homomorphism, *Topology* **8** (1969), 253-257.

[Be] A.A. Beilinson: Higher regulators and values of L-functions, *J. Soviet Mathematics* **30** (1985), 2036-2070.

[BK] A.A. Belavin,V.G. Knizhnik: Algebraic geometry and the geometry of quantum strings, *Physics Letters B* **168** (1986), 201-206.

[BGV] N. Berline, E. Getzler, and M. Vergne: *Heat kernels and the Dirac operator*, Grundlehren der Math. Wissenschaften **298**, Springer-Verlag (1992).

[BV] N. Berline and M. Vergne: A proof of Bismut local index theorem for a family of Dirac operators, *Topology* **26** (1987), 435-464.

[B1] J.-M. Bismut: The Atiyah-Singer index theorem for families of Dirac operators: two heat equation proofs, *Inventiones Math.* **83** (1986), 91-151.

[B2] J.-M. Bismut: Koszul complexes, harmonic oscillators and the Todd class, *Journal of the AMS* **3** (1990), 159-256.

[B3] J.-M. Bismut: Superconnection currents and complex immersions, *Inventiones Math.* **99** 1 (1990), 59-113.

[BF] J.-M. Bismut and D.S. Freed: The analysis of elliptic families I, Metrics and connections on determinant bundles, *Comm. Math. Physics* **106** (1986), 159-176.

[BGS1] J.-M. Bismut, H. Gillet, and C. Soulé: Analytic torsion and holomorphic determinant bundles I, II, III, *Comm. Math. Physics* **115** (1988), 49-78, 79-126, 301-351.

[BGS2] J.-M. Bismut, H. Gillet, and C. Soulé: Bott-Chern currents and complex immersions, *Duke Journal* **60** (1990), 255-284.

[BGS3] J.-M. Bismut, H. Gillet, and C. Soulé: Complex immersions and Arakelov geometry, *Grothendieck Festschrift* **I** (1990), Birkhäuser, 249-331.

[BKo] J.-M. Bismut and K. Köhler: Higher analytic torsion forms for direct images and anomaly formulas, *Journal Alg. Geometry* **1** (1992), 647-684.

[BL] J.-M. Bismut and G. Lebeau: Complex immersions and Quillen metrics, *Publications Math. IHES*, to appear.

[BVa] J.-M. Bismut and E. Vasserot: The asymptotics of the Ray-Singer analytic torsion associated with high powers of a positive line bundle, *Comm. Math. Physics* **125** (1989), 355-367.

[Bo] E. Bombieri: The Mordell conjecture revisited, *Ann. Sc. Norm. Sup. Pisa, Cl. Sci.*, IV **17** (1990), 615-640.

[BoGS] J.-B. Bost, H. Gillet, and C. Soulé: Heights of projective varieties and positive Green forms, *Journal AMS*, to appear.

[BC] R. Bott and S.S. Chern: Hermitian vector bundles and the equidistribution of the zeroes of their holomorphic sections, *Acta Math.* **114** (1965), 71-112.

[BT] R. Bott and L.W. Tu, *Differential Forms in Algebraic Topology*, Graduate Texts in Mathematics 82, Springer-Verlag (Berlin-Heidelberg) (1982).

[BG] K.S. Brown and S.M. Gersten: Algebraic $K$-theory as generalized cohomology, *Lecture Notes in Mathematics* **341**, Springer-Verlag (Berlin-Heidelberg) (1973), 266-292.

[CV] P. Cartier and A. Voros: Une nouvelle interprétation de la formule des traces de Selberg, *Grothendieck Festschrift* **II** (1990), Birkhäuser, 1-68.

[C] K. Chandrasekharan: *Arithmetical functions*, Grundlehren der Math. Wissenschaften **167**, Springer-Verlag (1970).

[Da] G. Dahlquist: On the analytic continuation of Eulerian products, *Arkiv för Math.* **1** (1951), 533-554.

[D] P. Deligne: Le déterminant de la cohomologie, Current Trends in Arithmetical Algebraic Geometry, *Contempory Math.* **67** (1987), 93-178.

[De1] C. Deninger: On the gamma factors attached to motives, *Inventiones Math.* **104**, 2 (1991), 245-262.

[De2] C. Deninger: Local L-factors of motives and regularized determinants, *Inventiones Math.*, **107** (1992), 137-150.

[DG] J. Dieudonné and A. Grothendieck: *Éléments de Géométrie Algébrique I*, Grundlehren der Math. Wiss. **166**, Springer-Verlag (Berlin-Heidelberg) (1971).

[DP] A. Dold and D. Puppe: Homologie nicht-additiver Funktoren. *Anwendungen, Ann. Inst. Fourier* **11** (1961), 201-312.

[Do] S.K. Donaldson: Anti-self-dual Yang-Mills connections over complex algebraic surfaces and stable vector bundles, *Proc. London Math. Soc.* **3**, 50 (1985), 1-26.

[DR1] G. De Rham: *Variétés différentiables: formes, courants, formes harmoniques*, Paris, Hermann (1960).

[DR2] G. De Rham: L'oeuvre d'Elie Cartan et la topologie, in *Oeuvres Mathématiques, L'enseignement Math.*(1981), 641-650.

[Dw] B. Dwork: On the zeta function of a hypersurface, *Publ. Math. IHES* **12** (1962), 5-68.

[E1] I. Efrat: Determinants of Laplacians on surfaces of finite volume, *Comm. Math. Physics* **119** (1988), 443-451.

[E2] I. Efrat: Eisenstein series and Cartan groups, *Illinois Journal Math.* **31**, 3 (1987), 428-437.

[El] R. Elkik: Métriques sur les fibrés d'intersection, *Duke Math. Journal* **61** (1990), 303-328.

[F1] G. Faltings: Endlichkeitssätze für abelsche Varietäten über Zahlkörpern, *Inventiones Math.* **73** (1983), 349-366.

[F2] G. Faltings: Calculus on arithmetic surfaces, *Annals of Math.* **119** (1984), 387-424.

[F3] G. Faltings: Diophantine approximation on abelian varieties, *Annals of Math.* **133** (1991), 549-576.

[F4] G. Faltings: *Lectures on the arithmetic Riemann-Roch theorem*, notes by S. Zhang, Annals of Math. Studies **127** (1992), Princeton University Press.

[Fu] W. Fulton: *Intersection Theory*, Ergebnisse der Mathematik und ihrer Grenzgebiete **3** Folge, Band 2, Springer-Verlag (Berlin-Heidelberg) (1984).

[Ger] S.M. Gersten: Problems about higher K-functors,in *Springer Lecture Notes* **341** (1973), 43-56.

[Ge] E. Getzler: A short proof of the local Atiyah-Singer index theorem, *Topology* **25** (1986), 111-117.

[Gi] P.B. Gilkey: *Invariance theory, the heat equation, and the Atiyah-Singer index theorem*, Math. Lecture Series **11** (1984), Publish or Perish.

[G1] H. Gillet: An introduction to higher dimensional Arakelov theory, *Contemporary Mathematics* **67** (1987), 209-228.

[G2] H. Gillet: Riemann-Roch for higher algebraic $K$-theory, *Advances in Math.* **40** (1981), 203-289.

[G3] H. Gillet: K-theory and intersection theory revisited, *K-theory* **1** (1987), 407-415.

[GS1] H. Gillet and C. Soulé: Intersection theory using Adams operations, *Inventiones Math.* **90** (1987), 243-278.

[GS2] H. Gillet and C. Soulé: Arithmetic intersection theory, *Publications Math. IHES* **72** (1990), 94-174.

[GS3] H. Gillet and C. Soulé: Characteristic classes for algebraic vector bundles with Hermitian metrics, I, II, *Annals of Math.* **131** (1990), 163-203 and 205-238.

[GS4] H. Gillet and C. Soulé: Amplitude arithmétique, *Note CRAS Paris* **307**, Série I (1988), 887-890.

[GS5] H. Gillet and C. Soulé: Analytic torsion and the arithmetic Todd genus, with an Appendix by D. Zagier, *Topology* **30**, 1 (1991), 21-54.

[GS6] H. Gillet and C. Soulé: On the number of lattice points in convex symmetric bodies and their duals, *Israel Journal of Math.*, **74**, 2-3 (1991), 347-357.

[GS7] H. Gillet and C. Soulé: Un théorème de Riemann–Roch–Grothendieck arithmétique, *Note CRAS Paris* **309**, I (1989), 929-932.

[GS8] H. Gillet and C. Soulé: An arithmetic Riemann-Roch theorem, *Inventiones Math.* **110** (1992), 473-543.

[GS9] H. Gillet and C. Soulé: Direct images of Hermitian holomorphic bundles, *Bull. AMS* **15** (1986), 209-212.

[Gr] P. Greiner: An asymptotic expansion for the heat equation, *Arch. Ration. Mech. Anal.* **41** (1971), 163-218.

[GH] P. Griffiths and J. Harris: *Principles of Algebraic Geometry*, John Wiley and Sons (1978).

[Gro] A. Grothendieck: Formule de Lefschetz et rationalité des fonctions L, *Séminaire Bourbaki* **279** , in *Dix exposés sur la cohomologie des schémas*, Masson, North-Holland (1968), 31-45.

[GBI] A. Grothendieck, P. Berthelot, and L. Illusie: SGA6, *Théorie des intersections et théorème de Riemann-Roch*, Springer Lecture Notes in Math. **225** (1971).

[GL] P.M. Gruber, C.G. Lekkerkerker: *Geometry of numbers*, North-Holland (1987).

[H] R. Hartshorne: *Algebraic Geometry*, Graduate Texts in Mathematics **52**, Springer-Verlag (Berlin-Heidelberg) (1977).

[Hi] H. Hironaka: Resolution of singularities of an algebraic variety over a field of characteristic zero: I, II, *Annals of Math.* **79** (1964), 109-326.

[Hz] F. Hirzebruch: *Topological methods in algebraic geometry*, Grundlehren der Math. Wissenschaften **131** (1966), Springer-Verlag.

[I] J. Itard: *Arithmétique et Théorie des Nombres*, (1967), P.U.F., Paris.

[K] C. Kassel: Le résidu non commutatif (d'après M.Wodzicki), *Séminaire Bourbaki* **199**, Astérisque 177-178 (1989), 199-230.

[Ki] M. Kim: Numerically positive line bundles on an Arakelov variety, *Duke Math. Journal* **61**, 3 (1990), 805-822.

[Kg] J. King: The currents defined by algebraic varieties, *Acta Math.* **127** (1971), 185-220 .

[Kg] J. King: Global residues and intersections on a complex manifold, *Trans. Amer. Math. Soc.* **192** (1974), 163-199.

[KT] S.L. Kleiman and A. Thorup: Intersection theory and enumerative geometry: a decade in review, §3, *Proc. of Symp. in Pure Math.* **46** (1987), 332-338.

[KM] F.F. Knudsen and D. Mumford: The projectivity of the moduli space of stable curves I: Preliminaries on "det" and "div", *Math. Scand.* **39** (1976), 19-55.

[Ko] K. Köhler: Holomorphic torsion on Hermitian symmetric spaces, (1994), preprint, Max-Planck-Institut für Math.

[Kr] C. Kratzer: Opérations d'Adams et représentations de groupes, *Enseign. Math.* **II** (1980), 141-154.

[Ku] N. Kurokawa: Analyticity of Dirichlet series over prime powers, in *Springer Lecture Notes* **1434** (1990), 168-177.

[Lf] L. Lafforgue: Une version en géométrie diophantienne du "Lemme de l'indice", Ecole Normale Supérieure, D.M.I.(1990), preprint.

[LW] E. Landau and A. Walfisz: Uber die Nichtfortsetzbarkeit einiger durch Dirichletsche Reihen definiertser Funktionen, *Rend. di Palermo* **44** (1919), 82-86.

[L] S. Lang: *Introduction to Arakelov theory* (1988), Springer-Verlag.

[Le] P. Lelong: Intégration sur un ensemble analytique complexe, *Bull. Soc. Math. France* **95** (1957), 239-262.

[Lv] H. Levine: A theorem on holomorphic mappings into complex projective space, Annals of Maths. **71**, 2 (1960), 529-535.

[Ma] H. Matsumura: *Commutative Algebra*, (1970) Benjamin.

[Mg] B. Malgrange: Division des distributions, *Séminaire Bourbaki* **203** (1960).

[Mi] J. Milnor: *Introduction to algebraic K-theory*, Annals of Math. Studies **72**, Princeton Univ. Press (1971).

[P] P. Philippon: Sur des hauteurs alternatives I, *Math. Ann.* **289** (1991), 255-283.

[Q1] D. Quillen: Higher Algebraic K-Theory I, *Lecture Notes in Mathematics* **341**, 85-147, Springer-Verlag (Berlin-Heidelberg) (1973).

[Q2] D. Quillen: Determinants of Cauchy–Riemann operators over a Riemann surface, *Funct. Anal. Appl.* (1985), 31-34.

[Q3] D. Quillen: Superconnections and the Chern character, *Topology* **24** (1985), 89-95.

[R] P. Ramond: *Field theory: a modern primer*, Frontiers of Physics **51**, Benjamin-Cummings (1981).

[RS] D.B. Ray and I.M. Singer: Analytic torsion for complex manifolds, *Annals of Math.* **98** (1973), 154-177.

[RJ] J. Roberts: Chow's moving lemma, in *Algebraic Geometry*, Oslo 1970, F. Oort (ed.), Wolters-Noordhoff Publ. (Groningen) (1972).

[RP] P. Roberts: On the vanishing of intersection multiplicities of perfect complexes, *Bull. AMS* **13** (1985), 127-130.

[Sa] P. Sarnak: Determinants of Laplacians, *Comm. Math. Physics* **110** (1987), 113-120.

[Se1] J.-P. Serre: Groupes de Grothendieck des schémas en groupes réductifs déployés, *Publ. Math. IHES* **34** (1968), 37-52.

[Se2] J.-P. Serre: *Algèbre locale, multiplicités*, Lecture notes in Mathematics **11**, Springer-Verlag (Berlin-Heidelberg) (1975).

[Sh] B.V. Shabat: *Distribution of values of holomorphic mappings*, Translations of mathematical monographs **61** American Math. Soc. (1985).

[S1] C. Soulé: Opérations en K-théorie algébrique, *Canadian Journal of Math.*, **37** 3 (1985), 488-550.

[S2] C. Soulé: Régulateurs, *Séminaire Bourbaki* **644**, *Astérisque* **133-134** (1986), 237-253.

[S3] C. Soulé: Géométrie d'Arakelov des surfaces arithmétiques, *Séminaire Bourbaki* **713**, *Astérisque* **177-178** (1989), 327-343.

[S4] C. Soulé: Géométrie d'Arakelov et théorie des nombres transcendants, in *Journées arithmétiques 1989*, Astérisque **198-200** (1991), 355-373.

[St] W. Stoll: A Casorati-Weierstrass theorem for Schubert zeroes of semi-ample holomorphic vector bundles, *Atti Accad. Naz. Lincei Me.Cl. Sci. Fis. Mat. Natur. Sez.* Ia (8) **15** (1978), 63-90.

[Sz] L. Szpiro: Degrés, intersections, hauteurs, *Astérisque* **127** (1985), 11-28.

[V1] P. Vojta: *Diophantine Approximations and Value Distribution Theory*, Lecture Notes in Mathematics **1239**, Springer-Verlag (Berlin-Heidelberg) (1987).

[V2] P. Vojta: Siegel's Theorem in the compact case, *Annals of Math.* **133** (1991) 509-548.

[V3] P. Vojta: Mordell's conjecture over function fields, *Inventiones Math.* **98** (1989) 115-138.

[Vo] A. Voros: Spectral functions, special functions and the Selberg zeta function, *Comm. Math. Physics* **111** (1987) 439-465.

[Wa] X. Wang: Thesis, Harvard (1992).

[W1] A. Weil: *Number theory, an approach through history, from Hammurapi to Legendre* (1984), Birkhäuser.

[W2] A. Weil: Sur l'analogie entre les corps de nombres algébriques et les corps de fonctions algébriques, [1939a] in *Oeuvres Scient.* **I**, 236-240, (1980), Springer-Verlag

[WW] E.T. Whittaker, G.N. Watson: *A course of modern analysis*, Cambridge University Press (1940).

[Z] S. Zhang: Positive line bundles on arithmetic surfaces, Thesis, Columbia (1990), to appear.

# Index